Architectural Technology up to the Scientific Revolution

New Liberal Arts Series

Light, Wind, and Structure: The Mystery of the Master Builders, by Robert Mark, 1990

The Age of Electronic Messages, by John G. Truxal, 1990

Medical Technology, Ethics, and Economics, by Joseph D. Bronzino, Vincent H. Smith, and Maurice L. Wade, 1990

Understanding Quantitative History, by Loren Haskins and Kirk Jeffrey, 1990

Personal Mathematics and Computing: Tools for the Liberal Arts, by Frank Wattenberg, 1990

Nuclear Choices: A Citizen's Guide to Nuclear Technology, by Richard Wolfson, 1991

Discovery, Innovation, and Risk: Case Studies in Science and Technology, by Newton Copp and Andrew Zanella, 1992

Architectural Technology up to the Scientific Revolution, edited by Robert Mark, 1993

This book is published as part of an Alfred P. Sloan Foundation program.

Architectural Technology up to the Scientific Revolution

The Art and Structure of Large-Scale Buildings

Robert Mark, editor

The MIT Press

Cambridge, Massachusetts

London, England

This book was set in Sabon by DEKR Corporation and was printed and bound in the United States of America.

Library of Congress Cataloging-in-Publication Data

Architectural technology up to the scientific revolution : the art and structure of large-scale buildings / edited by Robert Mark.
p. cm. — (New liberal arts series)
Includes bibliographical references and indexes.
ISBN 0-262-13287-7
1. Building—History. 2. Architecture—History. I. Mark, Robert. II. Series.
TH15.A73 1993
690'.094—dc20 92-35404
CIP

Contents

The Alfred P. Sloan Foundation's New Liberal Arts (NLA) Program stems from the belief that a liberal education for our time should involve undergraduates in meaningful experiences with technology and with quantitative approaches to problem solving in a wide range of subjects and fields. Students should understand not only the fundamental concepts of technology and how structures and machines function, but also the scientific and cultural settings within which engineers work, and the impacts (positive and negative) of technology on individuals and society. They should be much more comfortable than they are with making calculations, reasoning with numbers and symbols, and applying mathematical and physical models. These methods of learning about nature are increasingly important in more and more fields. They also underlie the process by which engineers create the technologies that exercise such vast influence over all our lives.

The program is closely associated with the names of Stephen White and James D. Koerner, both vice-presidents (retired) of the foundation. Mr. White wrote an internal memorandum in 1980 that led to the launching of the program two years later. In it he argued for quantitative reasoning and technology as "new" liberal arts, not as replacements for the liberal arts as customarily defined, but as liberating modes of thought needed for an understanding of the technological world in which we now live. Mr. Koerner administered the program for the foundation, successfully leading it through its crucial first four years.

The foundation's grants to 36 undergraduate colleges and 12 universities supported a large number of seminars, workshops, and symposia on topics in technology and applied mathematics. Many new courses were developed and existing courses modified. Some minors or concentrations in technology studies were organized. A Resource Center for the NLA Program, organized at Stony Brook, published a monthly newsletter and a series of monographs, collected educational materials prepared at the colleges and universities taking part in the program, and continues to serve as a clearing house for information about the NLA program.

As the program progressed, faculty members who had developed successful new liberal arts courses began to prepare textbooks. Also, a number of the foundation's grants to universities supported writing projects of professors—often from engineering departments—who had taught well-attended courses in technology and applied mathematics designed for liberal arts undergraduates. It seemed appropriate not only to encourage the preparation of books for such courses, but also to find a way to publish and thereby make available to the widest possible audience the best products of these teaching experiences and writing projects. This is the background with which the foundation approached The MIT Press and the McGraw-Hill Publishing Company about jointly publishing a series of books on the new liberal arts. Their enthusiastic response led to the launching of the New Liberal Arts Series.

The publishers and the Alfred P. Sloan Foundation express their appreciation to the members of the Editorial Advisory Board for the New Liberal Arts Series: John G. Truxal, Professor Emeritus, Department of Technology and Society, State University of New York, Stony Brook, Chairman; Joseph Bordogna, Professor and Dean Emeritus, School of Engineering and Applied Science, University of Pennsylvania; Robert W. Mann, Whitaker Professor of Biomedical Engineering, Massachusetts Institute of Technology; Merritt Roe Smith, Professor of the History of Technology, Massachusetts Institute of Technology; J. Ronald Spencer, Associate Academic Dean and Lecturer in History, Trinity College; and Allen B. Tucker, Jr., Professor of Computer Science, Bowdoin College.

In developing this new publication program, The MIT Press was represented by Frank P. Satlow, and the McGraw-Hill Publishing Company by Eric M. Munson. The first six books in the series were jointly published. Subsequent books have been solely published by The MIT Press, represented by Terry Ehling.

Samuel Goldberg
Alfred P. Sloan Foundation

Robert Mark
Editor

Sheila Bonde
Brown University

Lynn T. Courtenay
University of Wisconsin at Whitewater

Michael Davis
Mount Holyoke College

PETER FERGUSSON
Wellesley College

JOEL HERSCHMAN
Fordham University

CLARK MAINES
Wesleyan University

ROBERT MARK
Princeton University

ROWLAND RICHARDS, JR.
State University of New York at Buffalo

ELWIN C. ROBISON
Kent State University

ELIZABETH B. SMITH
Pennsylvania State University

LEONARD A. VAN GULICK
Lafayette College

For most teachers of architectural history who have studied traditional humanistic disciplines—history, art history, languages, and even architectural practice—the explication of historic building technology often requires a new "language" relating to such technological matters as the behavior of materials under load, techniques of construction, physics of natural lighting, and elements of modern structural analysis. Yet in spite of this difficulty, an understanding of the technological side of architecture is necessary to give students a more complete view of past attainments. This is especially important in the classroom where students, rather than merely noting the monuments, are predisposed to question the entire building process including such aspects as site conditions, availability and nature of materials, appropriateness of structure, and the wider social context of building and of patronage.

Although designed to be used as a "handbook" to aid both students and teachers in treating questions of technology in courses dealing with European architectural history before the Enlightenment, our intent is not to place technology in a privileged or a deterministic position. The full analysis of a building must take account of all the functions it was designed to meet, the iconographic and symbolic associations of its forms, the cultural contexts of its patrons and users, and a host of other issues. With our focus on technology, however, we hope to illuminate an aspect of the history of architecture that has too often remained obscure to our colleagues and students.

The text is the combined product of nine architectural historians (two of whom were trained originally as engineers) and two research engineers (with particular interests in historic building technology) who have worked together under the aegis of an Alfred P. Sloan Foundation–New Liberal Arts Program grant awarded to Robert Mark at Princeton. Its concept came from discussions that took place during the progression of a National Endowment for the Humanities-sponsored summer institute and several Sloan Foundation-sponsored seminars held at Princeton over the last five years that focused on the role of technology in historic architecture. The overall planning and final editing, which necessitated much interaction, was undertaken by the entire group, but principal authors of the chapters are listed in the table of contents. In addition to significant participation in both the writing and the editing of the text, Joel Herschman also assumed the chore of editing the photographs.

Though it stands alone, the text is not intended to supplant traditional, historical works; rather it deals with representative monuments from eras that witnessed the development of new, large-scale building types. Where appropriate, modern scientific tools are used to clarify the technological underpinning of these historically innovative structures and to provide insights into the design techniques employed by their builders. Starting with the subsoil below the buildings and proceeding upward to the apex of their timber roofs, introductory explanations of the typical structural components of historic buildings are offered in separate chapters: soils and foundations; masonry walls and other vertical elements, including piers, arches, and buttresses as well as some aspects of the physics of natural light (in order to deal with questions of wall openings); masonry domes and vaults; and timber roofs and spires. The most generally taught western monuments, including civil architecture ranging from the ancient Greek down to that of the Scientific Revolution, are then reexamined in light of this information, as diagrammed in table 1. A concluding chapter sums up the overall pattern of prescientific building design up to the Scientific Revolution and some of the reasons for its remarkable record of success. The circumstances that led first in Renaissance Italy, and then by the mid-seventeenth century virtually throughout Europe, to the displacement of the craftsman-master builder by the artist-architect are described along with a surprising consequence attributed to this transformation.

Table 1
Index to Historical Building Components

	Text Pages			
	Soils and foundations	Walls and other vertical elements	Vaults and domes	Timber roofs and spires
Introductory	16	52	138	182
Ancient	25	74	141	194
Imperial Rome	28	80	141	197
Byzantine	31	90	149	205
Early Medieval	34	98	153	207
Romanesque	35	100	153	210
Gothic	39	107	159	213
Renaissance	46	128	171	223

It is hoped that the book will gain attention also from general readers interested in architecture and architectural history. For this audience, as well as for students, professional jargon has been held to a minimum. The nature of the subject matter, however, demands that some technological ideas and terms be used. We have defined them as they appear throughout the text, in illustrations, and also in a glossary. All length and force dimensions are given in metric units. Hence it may be well to remind readers that 1 meter (m) equals 3.28 feet; 1 kilogram (kg) equals 2.2 pounds; 1 metric ton (T) equals 2,208 pounds or 1.1 English tons; a stress of 1 kilogram per square centimeter (kg/cm^2) equals 14.3 pounds per square inch (close to 1 atmosphere of pressure); a soil bearing pressure of 10 metric tons per square meter ($10\ ton/m^2$) is close to 1.0 English ton per square foot; and a material unit weight of 1,000 kilograms per cubic meter ($1{,}000\ kg/m^3$) is equivalent to 62.4 pounds per cubic foot (the unit weight of water).

The authors wish to express their gratitude to the Alfred P. Sloan Foundation and to the National Endowment for the Humanities for sponsoring the seminars and summer institute that sparked this endeavor. Text production was partially underwritten by an additional grant from the Alfred P. Sloan Foundation. The authors are most grateful also to Robert Bork and Andy Tallon, both of whom contributed far more to the enterprise than their formal title of "research assistant" may imply; to J. Weyman Williams for producing print conversions from the authors' color slides; and to Slobodan Ćurčić (Princeton University), Frank R. Horlbeck (University of Wisconsin), Walter Horn (University of California), Henrick M. Jansen (Svendborg & Omegns Museum), Robert Ousterhout (University of Illinois), and Gavin Simpson (University of Nottingham) for their generous contributions of illustrations. Finally, the splendid line drawings of Amy Stein deserve particular mention.

Architectural Technology up to the Scientific Revolution

Monumental architecture embodies a fusion of art, culture, social need, patronage, and technology. To explicate these facets of building, the historian normally synthesizes information from formal analysis, archaeological investigation, and the study of primary (written or visual) sources. Yet these methodologies alone do not permit definitive interpretations of large-scale building technology, particularly with regard to structure. A major problem is the paucity of pre-Renaissance writings on architecture (post-Renaissance architectural publication is discussed in chapter 6). For example, by far the most complete writing to come down to us about ancient construction is *The Ten Books of Architecture* by the Roman architect and military engineer Vitruvius.[1] *The Ten Books* tell us much about classical principles of architectural symmetry, harmony, and proportion as well as aspects of ancient construction including site selection, foundations, building materials, and even acoustics. Yet the work dates from the beginning of the imperial era, shortly before the birth of Christ and before the onset of the principal Roman building activity. Consequently, it contains no reference to the great amphitheaters and the domed and vaulted buildings that dominate our view of Roman architecture.

Similarly, the extraordinary sketchbook of Villard de Honnecourt, from the early thirteenth century, contains rudimentary building plans and elevations as well as several sketches and bits of text dealing with machines (including the water-powered sawmill illustrated in figure 1.1), applied geometry, and the setting out and cutting of stone. Yet the sketchbook gives few clues about the technical basis of design or of the materials used in building.[2] To deal with questions of Gothic construction, then, Villard needs to be interpreted in conjunction with more comprehensive yet still elusive, surviving lodgebooks from the fifteenth century.[3] Indeed, even the few extant records from meetings of specialists called upon to deal with questions of structural stability or preservation, including the so-called *Expertises* of Chartres (1316) and Milan (1401), are not entirely reliable. The Chartres *Expertise* is rather vague about problems with the then century-old cathedral's buttressing,[4] and the famous Milan *Expertise* suffers from being a summary of technical discussions by a scribe rather than a builder.[5]

A reading of Vitruvius and a sequence of writers on architecture well into the Renaissance leaves little doubt that Galileo's *Dialogues Concerning Two New Sciences,* published in 1638, some sixteen centuries after *The Ten Books,* constitutes the introduction of analytical structural mechanics.[6] In fact, the first science treated by Galileo in the *Dialogues* is not dynamics, for which the work is now far better known, but that branch of structural engineering designated as "strength of materials" (see figure 1.2). The kind of mathematically based predictive engineering we know today was unavailable before the time of Galileo and hence, the technology discussed in our book is the product of craft tradition.

Analysis of the geometry of architectural forms has long been used by scholars to define the evolution of styles, but it should be remembered that the application of simple geometric rules alone could never guarantee structural success. Primary forms such as the arcs of circles, triangles, rotated squares, etc.—particularly those that could simply be laid out by using common instruments such as the straightedge and compass—were used extensively by builders for their *conceptual* designs, but these relationships were modified or augmented whenever

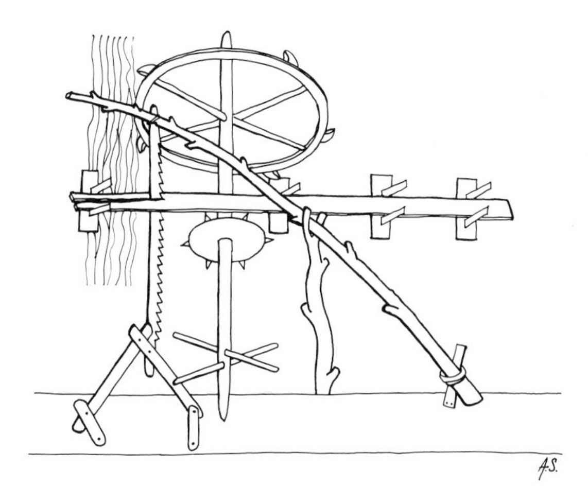

1.1 *Villard de Honnecourt: early thirteenth-century water-powered sawmill.*

1.2 *Cantilever beam in bending, from Galileo's* Two New Sciences, *1638.*

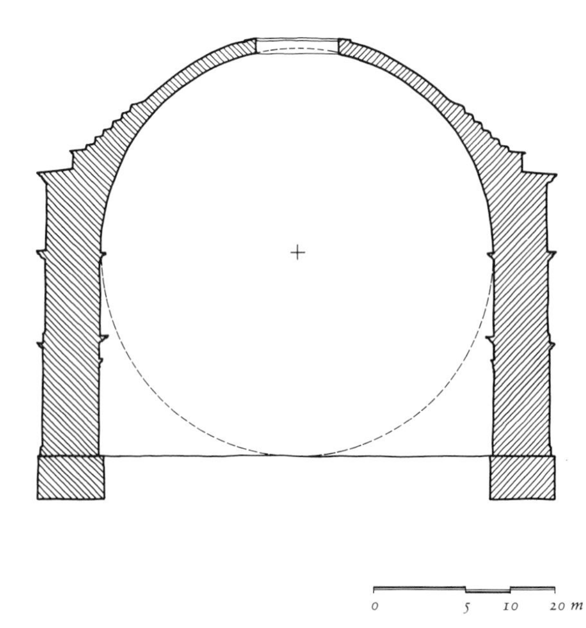

1.3 *Pantheon, Rome, A.D. 118–128: diagrammatic cross-section.*

necessary to secure structural stability. For example, although the interior space of the Roman Pantheon is bounded by a primary, hemispherical form, the extrados of the dome presents a much flatter, more complex profile (figure 1.3).

More significantly, the study of geometric forms does not take the *scale* of buildings into account. As it happens, this failing received paramount attention from Galileo because several of his contemporaries had attempted to define the (relative) strength of structures solely in terms of geometry.[7] Galileo thus opens the "First Day" of dialogues in the *Two New Sciences* with a description of activity in the Venetian arsenal:

Frequent experience of your famous arsenal, my Venetian friends, seems to me to open a large field to speculative minds for philosophizing, particularly in that area called mechanics. . . . Among its great number of artisans there must be some who, through observation handed down by their predecessors as well as those which they attentively and continually make of themselves, are truly expert and whose reasoning is of the finest. . . . [*These artisans*] *make the sustaining devices so much larger around that huge galley about to be launched than around smaller vessels,* [*and*] *this is done in order to avoid the peril of its splitting under the weight of its own vast bulk, a trouble to which smaller boats are not subject. . . . Now, all reasonings about mechanics have their foundations in geometry in which I do not see . . . large circles, triangles, cylinders, cones or any other figures or solids subject to properties different from those of small ones, . . .* [*yet*] *not only artificial machines and structures, but natural ones as well, have limits necessarily placed on them beyond which neither pure art nor nature can go while maintaining the same proportions and the same material.*[8]

1.4 Scaling.

The three-times larger-scale obelisk occupies a base nine times larger in area and encloses a volume (proportional to base area times height) 27 times larger than its smaller-scale counterpart. Since the pressure (or compressive stress) under the base is found by dividing the total weight of the structure by its base area, and if both structures are constructed of the same materials (or of materials having the same density), the larger structure will undergo three times as much compressive stress as that experienced by the similar, 1/3-scale structure.

In this particular case, if the obelisk is constructed of stone, the higher compressive stress in the larger structure would be expected to have little effect on the security of the obelisk itself, since stone is very strong in compression. But the greater compression acting on the soil below the larger obelisk could well prove critical to its stability (as discussed in chapter 2).

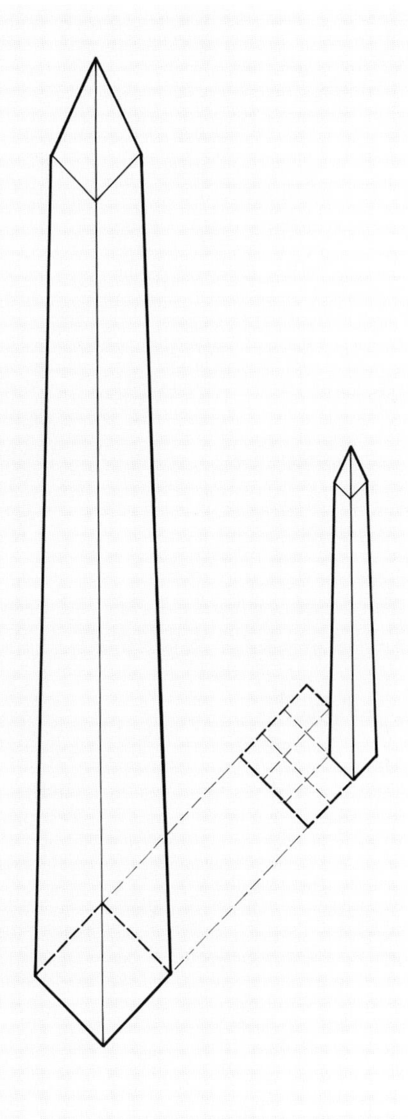

Galileo proves his thesis with a line of reasoning that may be briefly summarized by noting that a larger-scale building is subject to higher stresses because its *volume,* and therefore its dead weight, increases with the cube of building scale while the *area* of supporting structure within the building increases only as the square of its scale (see figure 1.4). Hence, geometry alone would never suffice to predict structural reliability, and one may further infer that any attempt to generate architectural form without regard to building *scale* is similarly fated.

The best evidence for elucidating the early builders' knowledge and working methods remains the buildings themselves. Indeed, the modern approach to questions of historic building construction derives mainly from archaeological investigations that accompanied nineteenth-century projects of restoration. Probably the most penetrating and useful single work to have originated under such circumstances is Eugene Viollet-le-Duc's profusely illustrated, ten-volume *Dictionnaire.*[9] Not surprisingly, Viollet-le-Duc's analyses serve to illustrate the interdependence of architectural form and the construction process, as well as the prodigious planning that underlay successful building projects. In our own time, archaeological inquiry has profited from using recently developed scientific methodologies that include radiocarbon, thermoluminescence, and dendrochronological dating techniques,[10] as well as physical and numerical/computer modeling to determine structural performance.[11]

The range of archaeological enquiry has also widened to encompass more of the building's vital structure, including regions normally invisible and therefore receiving scant attention. This holds true particularly for below-ground foundations, which for obvious reasons are critical to the initial planning phase of building as well as to the later well-being of any structure. In terms of quantity of construction material, it is not at all unusual to find as much stone placed below ground as above. New archaeological investigations and a reexamination of older data in light of modern analytical approaches for foundation design afford not only fresh insights into early building practice, but also new information that may lead to revising construction chronologies usually formulated without any consideration of foundations.

Also unseen in most historic architecture is the structural timber framing that supports the visually prominent roofs (for an example, see figure 1.5). If ventilated and kept dry, timber can endure for more than a millennium; hundreds of medieval timber roofs still survive, at least partially, in northern Europe. But roof structures from earlier eras in which the original timber work is no longer extant, as in ancient Greek and Roman buildings, must be (hypothetically) reconstructed by the historian to discern possible limitations placed on a particular building design by specifics such as maximum size of available timber. And of course, the ubiquitous timber centering that provided crucial support for both workman and stonework during the erection of historic masonry buildings disappeared long ago. As with foundations, new investigations and reexamination of archaeological data concerning roofs combined with techniques originally developed for modern structural analysis have helped interpret the master builders' intentions.[12] But before turning to

1.5 *Notre Dame Cathedral, Paris: southwest tower and nave roof, early thirteenth century.*

specific building components, it seems best to consider first some of the essential physical characteristics of building materials and basic elements of early design as well as how standards of measurement, so important in any communication between designer and builder, were managed in prescientific construction.

Building Materials

The most common materials used in early construction are timber, masonry (stone or fired brick), and unreinforced concrete. **Timber** differs from the others in its ability to accept tension (as anyone who has ever pulled at a young sapling can attest), although the tensile capacity of a timber structure is often limited by the strength of its joints rather than by the strength of the material itself. Hence timber strengths (in both tension and compression), which range from about 150 kg/cm^2 to 400 kg/cm^2 for both hardwoods and softwoods, can serve only as a guide for analysis. Both dimensional stability and the strength of timber are sensitive to changes in moisture content, and long-term loadings produce time-dependent deformation known as *creep*. Also, because of the natural growth variation of its internal structure, timber (even from a common source) is notorious for exhibiting a broad range of properties (usually due to knots and splits) and for being anisotropic; that is, displaying different structural responses depending on the direction of applied forces relative to the alignment of its grain.

Granite, limestone, and sandstone are the major constituents of **stone** masonry construction. Because of its durability, its relative workability, and its availability near most European building sites, limestone was the most commonly used material for load-bearing walls and piers. The strength of all stone in compression, but especially that of limestone, is variable. Like wood it is sensitive to such factors as the relationship between the stone's orientation in its natural bed (in situ) and the direction in which forces are applied to it in a building. The compressive strength of limestone ranges from a maximum of about 2,000 kg/cm^2 to a minimum of 200 kg/cm^2. Because of its brittleness (as with glass), the range of tensile strengths for all stone is an order of magnitude (i.e., ten times) lower than the range of compressive values. In turn, the tensile strength of stone is usually an order of magnitude greater than the tensile strength of the mortar used as grout between the surfaces of the blocks. To understand masonry performance, therefore, it is important to give attention also to the general properties of mortars.

Premodern **mortars** were generally composed of pure lime, or of lime and sand mixtures. Pure lime mortars are not hydraulic (that is, they do not set under water), and they are often described, erroneously, as taking anywhere from months to centuries to dry. In fact, lime mortar passes through two separate stages: setting and carbonation. Pure lime mortar paste is said to have "set" when all the excess water has evaporated into the atmosphere or has been absorbed into the surrounding masonry blocks. The time taken by this process varies with the amount of excess water, the relative humidity, the absorption rate of the stone, and the mass of the construction. Lime mortars are described as "slow-setting" only in contrast with modern Portland cement, which with a (chemical) accelerator sets in about ten hours. Lime mortar may take days or perhaps weeks to set, but not years. Carbonation, by comparison, is a much slower chemical process. The set mortar paste, calcium hydroxide, reacts with carbon dioxide to form calcium carbonate, the basic constituent of limestone. Under ordinary circumstances, the process is slow because of the small amount of carbon dioxide available in the atmosphere, and diffusion of carbon dioxide beyond a thin surface layer of carbonated mortar deep into the masonry joint occurs at a much slower rate, if at all. Although lime mortar that is only set is not very strong even in compression, neither the strength of the mortar nor that of the masonry blocks is as important as the properties of their combined construction.

The strength and deformation characteristics of masonry construction are difficult to predict. Even test results on discrete samples of materials are not accurate indicators of the behavior of the same materials when used in large quantities in buildings. Studies of masonry walls loaded perpendicular to the mortar bed, for example, have shown that mortar can survive under conditions in which its simple crushing strength is exceeded by as much as 300 percent.[13]

Unlike the simple lime mortars produced by adding water to a mixture of quicklime and sand, cements consisting of dark volcanic sand and lime—or as in today's Portland cement, of limestone and clays or shale burned in a furnace and then pulverized—set by combining chemically with water. These cements are hydraulic. Since they do not need to dry out, large batches of hydraulic cement will (chemically) cure relatively rapidly even in damp conditions. Thus, Roman pozzolan cement (named after the town of Pozzuoli, near Naples, where the volcanic ash used in its manufacture was discovered) could be used for the massive concrete structural elements of large buildings as well as for underwater construction. The early compressive strengths of hydraulic cements are also far superior to those of lime mortars. Recent studies have indicated that one must remain cautious, however, in characterizing the resistance of any unreinforced concrete—whether used as mortar within masonry or as a solid mass—to cracking caused by tensile forces.[14] Although modern concrete made with controlled-cured, high-quality Portland cement exhibits measurable tensile strength, in modern reinforced-concrete design its tensile capacity is still taken to be nil. Experience has dictated that reinforcing steel is always needed in the regions of a concrete structure where tension is present.

The substitution of kiln-fired **brick** for stone in masonry construction, other things being equal, makes little difference in overall structural behavior. On-site construction, however, would have been simplified and costs reduced by employing smaller units of a more uniform building product than was provided by preindustrial finished stone.

Although far less commonly employed than timber and masonry in building construction before the Scientific Revolution, **metals** possess a wide variety of useful properties that have made them indispensable to builders since the earliest periods of civilization. In particular, **wrought iron**—extracted from molten ore by burning away the carbon in repeated heating cycles in a charcoal fire, hammering and quenching with water, and then forging, or hammering into the desired shape—proved attractive to early builders because of its malleability and high tensile strength (up to 5,000 kg/cm^2). Though costly throughout antiquity and the Middle Ages, wrought iron was continuously available in modest quantities. Applied mainly in inserts designed to counteract the effect of forces that might cause masonry to shift, such as cramps tying blocks of stone together or as temporary ties during construction before the mortar had fully set, iron was seldom utilized for primary structural members—even after its application became increasingly frequent between the tenth and thirteenth centuries.[15]

The greater availability and quality of iron and other metals played an important part in the architecture of the late Middle Ages above and beyond purely structural concerns. The improved hardness of carbon-steel tools, for example, allowed the carving of hard stone with great precision.[16] In fact, developments in metallurgy were necessary prerequisites for the virtuoso performances of the late Gothic masons. Similarly, the flowering of the stained-glass window was facilitated by the use of iron armatures hung between the tracery to bear the weight of the glass panels, each of which was itself held in place by **lead** caulking.[17] Finally, the roofs of the great cathedrals and other large, permanent structures were often sheathed in lead.

Mensuration and Design

Early monumental building plans seem to have been largely based on proportion and geometry. Dimensions for fourth-century B.C. Greek temples, for example, were typically given as fractions or multiples of a modular column diameter.[18] The architect might provide a model of the temple corner, and full-scale dimensions could then be determined from measurements taken from the model and multiplied by an appropriate scale factor—without dependance on standard dimensional units. By the first century A.D., the pragmatic Roman bureaucracy instituted standardized measurements throughout the empire, allowing architects to employ specific units of length. Roman units of length, probably derived from much earlier use of typical dimensions of human extremities, and designated as such, were the *pes* (foot), *palmus* (hand), and *digitus* (finger). The length of the pes, equal to 4 palmi, or 16 digiti, could not have been established with modern precision, but it has been generally estimated as being equivalent to 0.295 meters.[19] (For comparison, a modern English foot converts to 0.3048 m.) After the fourth-century transfer of the political center of the empire to Constantinople, the standard of measure evidently became larger; the so-called Byzantine foot is equivalent to 0.312 meters.[20]

With the collapse of the empire in the west in the fifth century, the different regions of Europe and even cities close together in the same region adopted their own standards of measurement. These often retained the Roman designations for units, but standard lengths could differ by as much as 70 percent.[21] One consequence of this jumble of standards was that medieval master masons were known to carry with them **measuring rods** of wood that were divided into commonly used proportional ratios (1/2, 1/3, 1/4, etc.). In effect it was these rods that provided the modules—used in conjunction with geometric constructions derived from manipulating simple geometric figures such as squares and triangles (figure 1.6) with basic instruments such as the level, square, triangle, compass, and straightedge—that produced the plans of the great cathedrals.[22]

To translate these constructions to actual building, designs were often incised at full scale on flat surfaces, or **tracing floors.** Extant engravings of this type, evident still in the floors of some buildings (as illustrated in figure 1.7) and on vertical walls, encompass the entire building process from prelim-

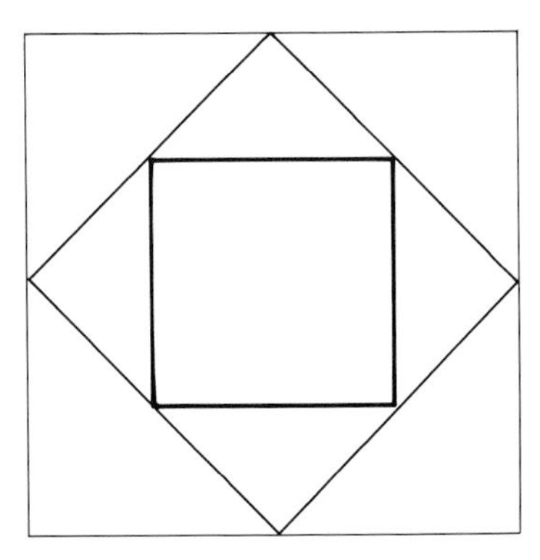

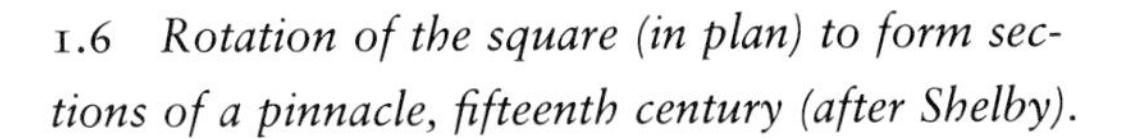

1.6 *Rotation of the square (in plan) to form sections of a pinnacle, fifteenth century (after Shelby).*

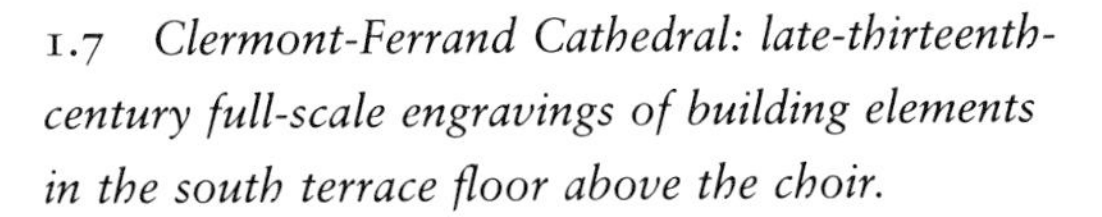

1.7 *Clermont-Ferrand Cathedral: late-thirteenth-century full-scale engravings of building elements in the south terrace floor above the choir.*

inary and incomplete sketches of such elements as ground plans or pier sections to finely finished depictions of windows, gables, and flying buttresses. These engravings seem also to have allowed the master mason to evaluate the aesthetic merits of a design against the rising structure, and their occasional abandonment demonstrates that significant modifications were effected during the late stages of execution. Incised window tracery plans would no doubt have been used to guide the glaziers, and the inclusion of masonry joints in some of the plans further suggests that they supplied information to the quarry—a procedure intended to economize material, labor, and transportation costs. For the cutting of complex stone elements, the engravings would necessarily have been supplemented by more easily handled templates.

Despite the thoroughness of these procedures and the obvious care in mensuration with which many of them were produced, it would have been difficult, if not impossible, to maintain a high degree of accuracy over long distances or high elevations, because cordage would have been used for such measurements. Unlike modern tapes manufactured from special stable alloys, cord lengths are sensitive both to the amount of tension present and to environmental variations such as humidity and temperature.[23]

Where reliance was made on units of measure, great pains were taken to simplify numerical calculation. Much of the famous discussion regarding the form of Milan Cathedral, for example, centered on providing a design in which the principal building elements could be easily laid out using simple whole number units to avoid complicated fractions (see fn. 5). By the sixteenth century, however,

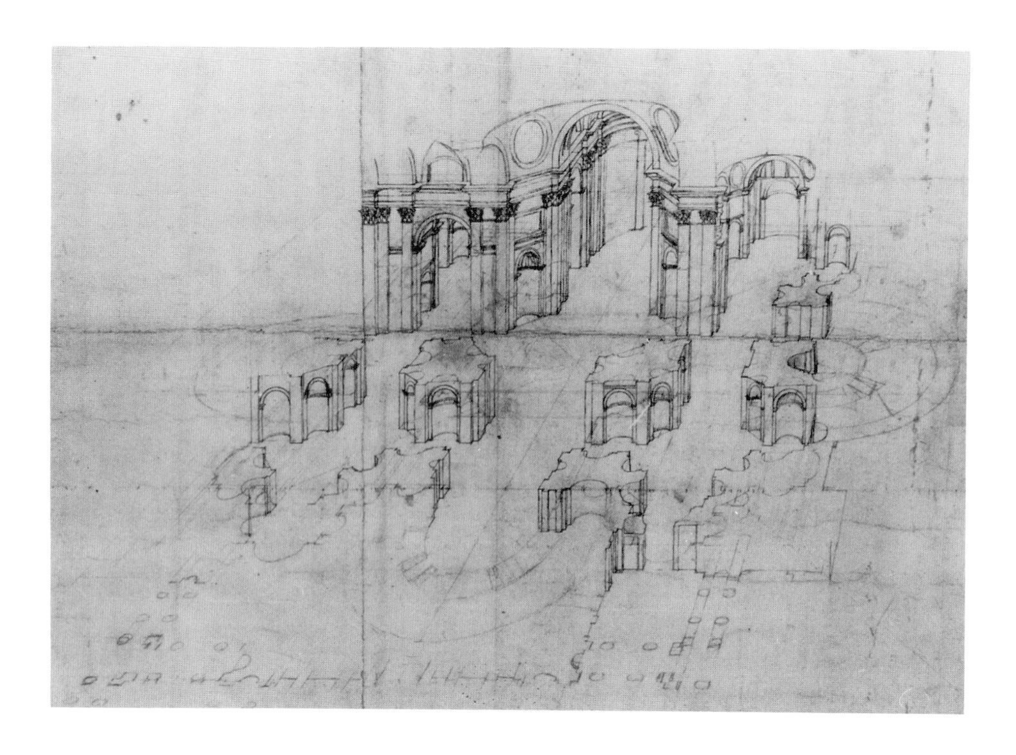

1.8 *Perspective view of the structure of St. Peter's Basilica, Rome, after Baldassare Peruzzi, 1529.*

most regions of Europe had well-defined standards of measurement as well as laborers with the necessary arithmetic skills to deal with numerical units, thus decreasing architects' reliance on geometric constructions in building design. For example, the Italian architect Palladio gives the dimensions of his engraved plans in his treatise in his local unit of measurement, the Vicentine foot (0.347 m). These dimensions are not related to simple geometric figures, but rather to numeric ratios that correspond to defined proportions, in some cases based on musical harmonics.[24] Although each region of Europe had its own units of measurement, by the sixteenth century there were accepted standards upon which architects could rely.

Another important later development for the communication of a design to the builders was the standardization of the three basic **orthogonal projections:** plan, elevation, and section. A plan shows a slice of the building cut along a horizontal plane, while both elevations and sections show slices taken at right angles through a vertical plane. The interrelationship of these drawings, to a common scale, is a practice that first became standard in the design of St. Peter's in Rome early in the sixteenth century. Indeed, Peruzzi's sectional perspective of St. Peter's (figure 1.8) could not have been formulated without these standardized relationships between orthogonal projections. Although models were, and are today, used to help visualize complicated spaces, the plan, section, and elevation drawings quickly became the primary tool used to define the relationship between the building plan and its superstructure.

Bearing all of this in mind, we may now go on to a sequential discussion of the structural components found in large-scale historic buildings.

Notes

1. Vitruvius, *De architectura*; trans. M. H. Morgan, *The Ten Books of Architecture* (New York, 1960).

2. Carl F. Barnes, Jr. *Villard de Honnecourt, The Artist and his Drawings: A Critical Biography* (Boston, 1982).

3. L. Shelby, *Gothic Design Techniques* (Carbondale, 1977).

4. Alan Borg and R. Mark, "Chartres Cathedral: A Reinterpretation of its Structure," *Art Bulletin,* vol. LV (September 1973), pp. 367–372.

5. According to John White (*Art and Architecture in Italy 1250–1400,* 2nd ed., Harmondsworth, 1987, p 521), the *Expertise* consists of "over-simplified notarial summaries of long and complex arguments . . . written in very crude and occasionally uncomprehending Latin." See also: J. S. Ackerman, "'Ars sine scientia nihil est'; Gothic Theory of Architecture at the Cathedral of Milan," *Art Bulletin* XXXI, no. 2 (June 1949), pp. 84–111.

6. Galileo Galilei, *Dialogues Concerning Two New Sciences* (1638), trans. S. Drake (Madison, 1974).

7. An idea not unlike the search by Renaissance architects for an aesthetic founded on a system parallel to the arithmetical principles of musical harmony uncovered by the ancients; see Rudolf Wittkower, *Architectural Principles in the Age of Humanism* (New York, 1971), pp. 101–154.

8. Galileo, *Dialogues* (1974), pp. 11–13.

9. E. E. Viollet-le-Duc, *Dictionnaire raisonné de l'architecture française du XIe au XVIe siècle,* 10 vols. (Paris, 1854–1868).

10. See Robert Berger, ed., *Scientific Methods in Medieval Archaeology* (Berkeley, 1970).

11. See Robert Mark, *Light, Wind, and Structure* (Cambridge/New York, 1990).

12. See, for example, Lynn T. Courtenay, and R. Mark, "The Westminster Hall Roof: A Historiographic and Structural Study," *Journal of the Society of Architectural Historians,* 46 (1987), pp. 374–393.

13. Sven Sahlin, *Structural Masonry* (Englewood Cliffs, NJ: 1971), pp. 52–56.

14. Mark, *Light, Wind, and Structure,* pp. 66–67.

15. See D. Knoop, and G. P. Jones, *The Medieval Mason,* 3rd ed. (New York, 1967), pp. 56–57.

16. Hard, carbon steel, also prized for weaponry, was produced by packing wrought iron with a carbonaceous material such as charcoal and reheating the assembly before a final quenching (carbon steel contains from 0.12 to 1.7 percent carbon: wrought iron contains lesser amounts). See Robert J. Forbes, *Studies in Ancient Technology* (Leiden, 1964), p. 31ff.

17. See Viollet-le-Duc, *Dictionnaire,* Vol. VI, pp. 317–345.

18. J. J. Coulton, *Ancient Greek Architects at Work* (Ithaca, 1977), pp. 57ff.

19. William L. Macdonald, *The Architecture of the Roman Empire,* I (New Haven, 1982), p. 140; see also William B. Parsons, *Engineers and Engineering in the Renaissance* (Cambridge, 1967), p. 626.

20. Rowland J. Mainstone, *Hagia Sophia: Architecture, Structure and Liturgy of Justinian's Great Church* (New York, 1988), p. 177.

21. Parsons, *Engineers and Engineering,* p. 631.

22. Shelby, *Gothic Design,* pp. 62ff.

23. P. Frankl, *The Gothic: Literary Sources and Interpretations Throughout Eight Centuries* (Princeton, 1960), pp. 66, 136.

24. Wittkower, *Architectural Principles,* loc. sit.

2.1 *Durham Cathedral, 1093–ca. 1137, sited on projecting bedrock.*

All soils consist of solid grains of different size and shape, and of voids filled with air and/or water. **Soil properties** are primarily related to the size of the soil particles, with the predominant types, *sand, silt,* and *clay,* displaying descending orders of grain size. The air between the soil particles is of little engineering significance and can be ignored. Changes in water content and the flow of water through soils, on the other hand, can be very important. Fluctuations in the height of the groundwater table may cause swelling or settlement in clays as water is either absorbed or squeezed out of the voids. The rate at which water flows through soil (in a standard soil test), known as *permeability,* is highly dependent on the size and shape of the soil grains. Short-term effects resulting from seasonal precipitation patterns or changes in drainage systems near structures can be crucial factors affecting the stability of buildings constructed on silts, but are less critical for buildings constructed on clays through which water migrates more slowly. Long-term changes in ground water levels, however, such as the effect of wells serving growing populations, can result in damaging settlement, as recent restorations at Winchester and York Cathedrals demonstrate (Bussby 1987, 1979, 56; Phillips 1976, 260).

The introductory New Testament passage goes on to contrast the sensible man who builds on rock with the foolish man "who built his house on sand. Rain came down, floods rose, gales blew and struck that house and it fell; and what a fall it had" (Matthew 7:27). Yet coarse sand is actually an excellent foundation substrate, at least for deep foundations. It is easily compacted and drains well, and it displays good *shear strength* (resistance to sliding between grains) when adequately confined. The

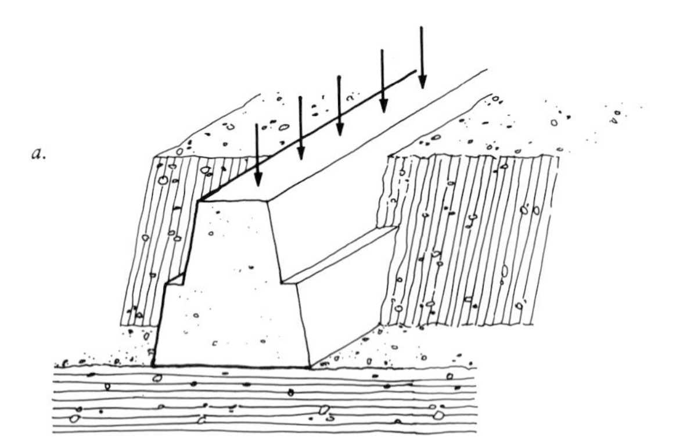

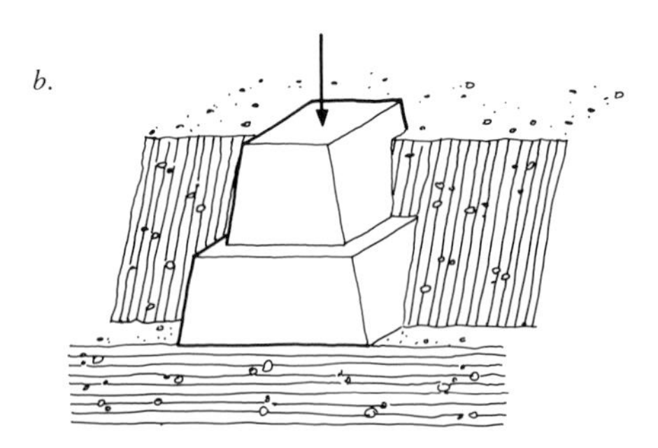

2.2 *Footing types: (a) continuous-wall footing; (b) isolated, single-column footing.*

strength of unconfined soil in shear is simply related to the maximum angle taken by the sides of an excavation before the slope fails and fills the hole. For example, to maintain stability, a hole dug in dry sand must have shallow sides, while stiffer and more cohesive soils, such as clay, can be excavated with perpendicular, vertical sides. The low inherent shear strength of unconfined sands is offset, however, by digging a deep foundation so that the tendency of sand to spread outward and bulge upward around the footing is countered by the confining weight of the surrounding soil. Thus the bearing capacity of sandy soils depends almost entirely on the depth of the foundation. When confined, sands have allowable bearing capacities of 20–40 metric tons/meter^2, about twice that of most clays (cf. table 2.1). When used as a load-bearing material, mortar and cobbles behave much like sand in that there is little natural shear strength, but it too becomes strong when confined by surrounding soil or a wall.

Clays exhibit a wide range of properties, depending on grain size, grain shape, and chemical composition. Often they have even poorer resistance to shear than sands (a condition experienced in a slippery mud puddle), and they expand and contract upon wetting and drying. Clays also tend to change volume when external loads are applied and may creep (continue to deform over time under loading) for many years. For sand, which has much higher permeability than clay, volume changes due to changes in water content are slight and occur rapidly. In clay, where the volume of voids is large and the water can escape only slowly through the pores, settlements may be large and can take decades to occur. Hence, foundations built on previously compressed clay soil (such as when a new building replaces another on the same site) are less likely to exhibit significant long-time deformation.

Because clay particles are so small, water can also be bound in the soil structure on a molecular level. If the clay was deposited in a salt-water condition, as occurred when oceans covered much of the land mass in early geologic periods, the introduction of fresh water can result in surprisingly large (and damaging) expansion and shrinking. Fortunately, these expansive clays are not commonly found in Europe (nor in eastern North America). As with sand, clay soils adjacent to the contact surface below a foundation exert downward and lateral pressures by virtue of their weight, which helps to prevent the compressed soil directly under the foundation from bulging outward and upward. For fine clays, however, this overburden effect is small in

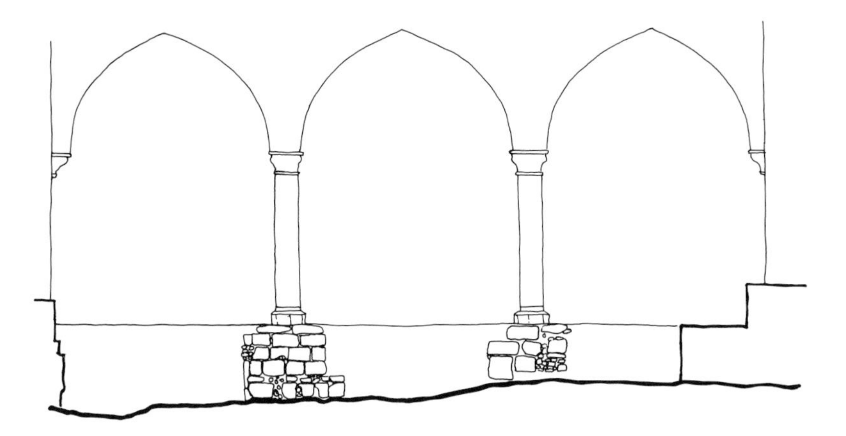

2.3 *Abbey of Saint-Jean-des-Vignes, Soissons: cross-section showing foundations and site geology for the Gothic chapter room in the eastern claustral range, first half of the thirteenth century (wall shown above the column bases is a reconstruction).*

comparison to their natural cohesive strength. Thus, confinement provided by greater foundation depth is less important for clay than for sandy soils; depending on the subsoil type, deep foundations do not in themselves indicate superior building skill.

As might be expected, silts exhibit properties that range between those of sands and clays, depending on grain size and shape. Larger-grained silts behave much like sands, while smaller-grained silts can approach clays in physical properties, making the determination of the bearing capacities of silts much more difficult than for sand. Indeed, builders could only have predicted behavior by comparison with other buildings sited on the same soil stratum.

Discrete foundations are usually one of two basic **footing types:** *continuous wall* and *single column,* as illustrated in figure 2.2. To support the total weight of aboveground walls, vaults, and roofs, as well as the weight of the footings themselves (minus the weight of the displaced soil), both types of footings must provide sufficient contact area so that the bearing capacity of the soil below them is not exceeded. Although early builders could have estimated the volumes and perhaps also the weight of buildings (for volumetric estimates were necessary to determine quantities of building materials and to pay for completed work), they would have been unable to quantify a given soil's bearing capacity. Their success depended on their empirical understanding gained from building experience. For example, the thirteenth-century builders of the eastern range of the cloister at the abbey of Saint-Jean-des-Vignes in Soissons dug trenches for continuous wall footings to the same geologic layer for the structure's two outer walls, even though footings supporting the same weight descended to a different depth on opposite sides of the same building (figure 2.3). Presumably the builders believed, on the basis of practical experience, that the layer on which they set the footings was the one needed to support the building adequately, regardless of whether they had to go deeper on one side of the building than the other to reach it. In modern engineering practice it is (conservatively) assumed that the load spreads outward through the soil at a slope of two downward, vertical units to each horizontal unit (figure 2.4). Thus for a continuous wall footing, the soil pressure decreases inversely with the depth beneath the interface, and the presence of weak soil layers far below footings is normally of little consequence.

2.4 Soil pressure distribution below a footing.
The breadth of the "supporting area" of soil below a continuous wall footing *increases directly with depth. The total breadth of the supporting area, therefore, is equal to the sum of the footing width plus the depth (as measured from the bottom surface of the footing); for example, the theoretical bearing pressure at a depth below the footing equal to the footing width (b) is but 1/2 the bearing pressure at the base of the footing.*

For an isolated, single column footing, *where the boundaries of the supporting area of soil below the footing increase in all directions with depth, the supporting area at a depth below the footing equal to the footing width is four times the footing base area; hence, the theoretical bearing pressure there is but 1/4 that at the base of the footing.*

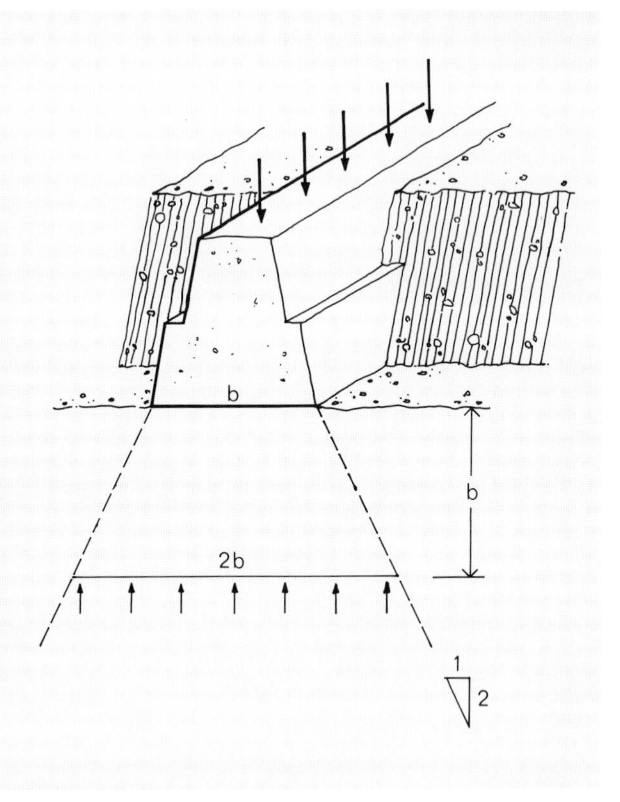

The frost line must also be taken into consideration when determining foundation depth. Frost heaves occur when a footing is placed on soil in which water freezes and expands, causing the volume of the supporting ground to change. Water expands approximately 10 percent by volume when it freezes, and the forces generated by contained ice can be extremely large (on the order of 2,000 kg/cm^2). Chalky or silty soils are most likely to heave when frozen, sands being much more stable due to their greater permeability. In most of western Europe, a depth of about one-half meter (from the soil surface to the top of the footing) is today generally considered safe to avoid frost heave, whereas in the northern United States and southern Canada, the colder winters require footing depths of over a meter.

Differential **settlement** (as opposed to the general, uniform settlement of all parts of a building) is most critical, particularly for masonry construction. For example, when it became the fashion around the beginning of the fourteenth century to crown English churches with great central towers and spires, their crossing piers often sank relative to the adjacent structure, leading to grievous cracking of the building fabric. At Ely Cathedral, poor soil conditions figured prominently in the collapse of the

2.5 *Ely Cathedral, nave, mid-twelfth century. Note a sag in the wall near the northern facade tower, the effect of differential settlement.*

northern tower of the facade as well as the large central tower over the crossing, later replaced by the remarkable light, timber Octagon. The magnitude of differential settlement between the heavy facade towers and the lighter nave, which caused the collapse, is still visible at the junction between the nave and the western end (figure 2.5).

Differential settlement may occur in new construction as well. To avoid this problem, the bearing pressure applied by the footing to a (consistent) soil stratum should be equal at all points under the structure. Heavier loads, such as those associated with towers, require larger footings to equalize bearing pressure, and there is evidence to suggest that some early builders understood this requirement. For example, the foundations beneath the central vessel piers of Meaux Cathedral are deeper and wider than those of the side aisle walls (Kurmann 1971, 70). While this condition may have resulted from the fourteenth-century partial rebuilding of the twelfth-century cathedral, it nevertheless reflects an instance of asymmetrical foundations. Excavations within the westernmost bays of medieval churches would probably reveal that towers were often more broadly footed than the rest of the building. Conversely, footing sizes should be adjusted to match the different soil conditions present on the site. Large public buildings, like Gothic cathedrals, were often laid across soils of different bearing capacities or were set partially upon preexisting structures, whose walls and/or foundations were reused by the newer building. Avoiding problems of differential settlement resulting from these conditions requires far better theoretical understanding than does building design on uniform "virgin" soil, and such knowledge would have been unavailable. Yet in many of these projects, the great weight of the volume of material used for new foundations, together with the relatively long period of time that typically elapsed between the setting of the foundations and the carrying out of appreciable aboveground construction, could well have produced much of the final differential settlement *before* the building superstructure was put into place. Differential settlement in the building itself would thus have been minimized.

Catastrophic failure of an individual footing, or of an entire building, occurs when the ultimate bearing capacity of a soil is exceeded. The so-called Leaning Tower adjoining the cathedral at Pisa, built between 1174 and 1370, is the best-known medieval example of this type of catastrophic failure (in "slow motion"). Soil creep below its shallow foundation produces a continuing tilt that has brought the top of the tower over four meters out of true vertical (figure 2.6). More commonly, catastrophic failure of foundations occurs when large settlements affect the stability of arches or vaults (see chapter 3). Yet our knowledge of many early failures is scant. Early writers seldom commented upon, much less understood, reasons for foundation failure. A notable exception has come down to us in the Miracles of Saint Benedict, which describes "the inadequacies of the foundation provoked early in its life the collapse of the northern part" of the vaulted transept of a priory at Perrecy-les-Forges. Moreover, the text describes how in c. 1095, the monks "reinforced the piers of the northern part," providing a rare glimpse of a foundation problem and its correction in the eleventh century.

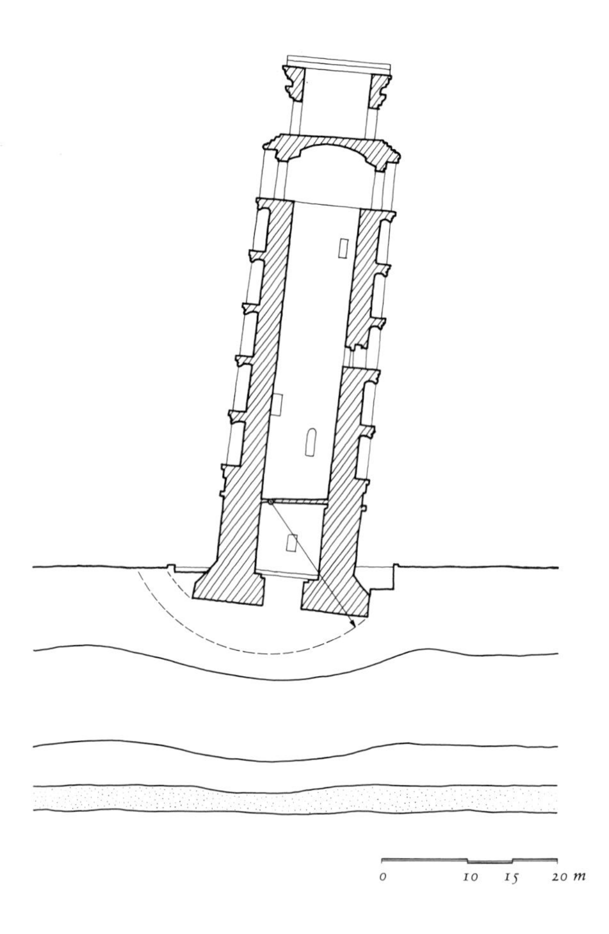

2.6 *Campanile, Pisa Cathedral, begun 1174: soil layers below and the effective failure slip-surface. Excessive settlement and rotation, due to low shear strength of the supporting soil, have characterized the "leaning tower" since the thirteenth century (after Kerisel).*

When discussed in the modern literature, foundations are usually characterized simply as the subterranean extensions of aboveground forms. Nevertheless, closer examination of the foundations of buildings from ancient Greece to Renaissance Europe reveals the full extent of the subtle dialogue between planning, technology, and structural art. If the dome of the Pantheon or the vaults of Amiens Cathedral continue to fill the modern observer with awe, it is in no small measure due to the ingenuity of their designers in matching foundations to the particular soil conditions at their sites.

Ancient

The history of Greek temple architecture is traditionally organized by reference to the evolution of capital types and column proportions. Temples of different orders, however, all distribute the loads concentrated on their columns to substructures that are remarkably similar. Rows of columns usually rested on orthostats, long blocks of stone laid horizontally (figure 2.7). Below the orthostats, foundations were normally stepped outward to include two substructure courses. Like aboveground walls, these courses were normally built without mortar and held together with iron clamps (see chapter 3). Stepped foundation walls of a nearly vertical profile were usually placed under the cella wall and on the interior face of the colonnade. The exterior face under the stylobate was, however, often more widely splayed. Where possible, it seems that Greek builders

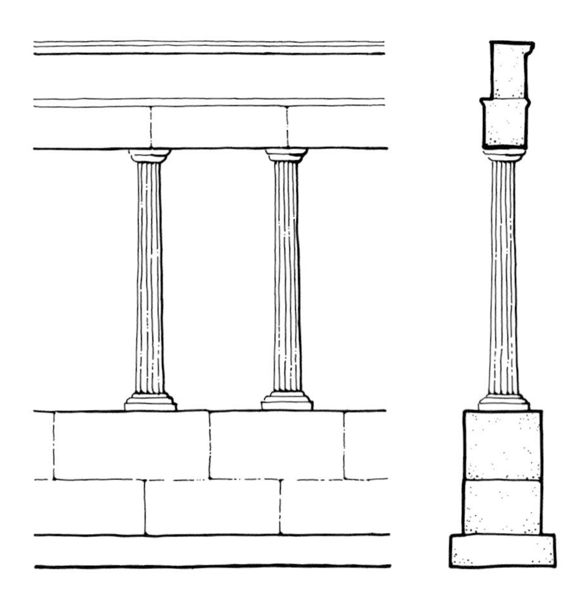

2.7 *Doric order: underground (orthostats) and supporting structure above (after Kerisel).*

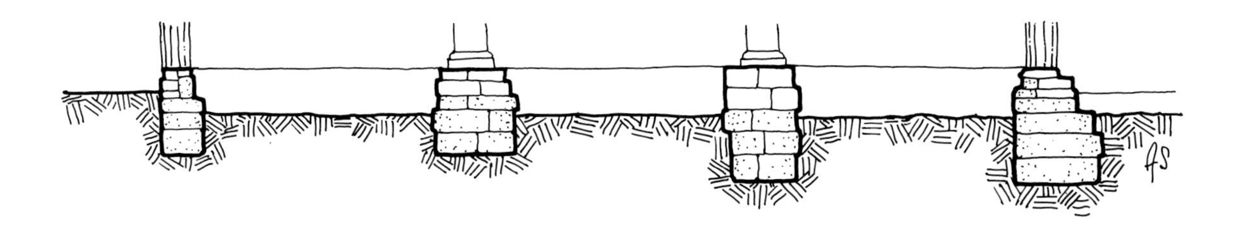

2.8 *Delos, third century B.C., porticus. Section showing isolated, single column-type footing used below individual piers (after Martin).*

2.9 *Temple of Apollo, Delphi, ca. 513–505 B.C.: foundation grid.*

sunk their foundations to bedrock, which tended to be relatively close to the surface. This is confirmed by textual references, with ancient authors prescribing bedrock or virgin soil (Martin, 310). Although continuous foundation walls seem to have been the rule, there were exceptions. Isolated pier footings are found under inner colonnades as well as in hypostyle construction. A good example is provided by the porticus at Delos (figure 2.8), where individual footings descend to the variable depths of the natural bedrock.

The site of Greek public buildings and temples was usually chosen with great care. A hilltop location was preferred for temples, while stoas often faced south to provide a sunny and sheltered spot for the many public activities that took place within them. Because little of the ground in Greece is naturally flat, building sites were often leveled, either by artificial terracing or by cutting into hillsides. The Bouleterion (Senate House) in Athens, for example, was set on a flat area created by quarrying out the bedrock at the base of the hill. The substructures of the Temple of Apollo at Delphi reveal an effective solution to a problematic sloping site (figure 2.9). Here the foundation takes the form of a grid constructed from uniformly sized ashlar blocks (Mainstone, 176).

Breccia, or conglomerate stone, was often reserved for foundations because it could not be readily dressed smooth for aboveground construction. The Temple of Apollo at Didyma, for example, employs foundations of roughly hewn conglomerate to support fine ashlar walls (figure 2.10). The hierarchy of building material is particularly apparent at

2.10 *Temple of Apollo, Didyma, ca. 330 B.C. Rough-conglomerate foundations.*

Temple "A" at Cos, where the porous stone of the foundations is separated from the marble of the upper columns and walls by a limestone course. Though cement and mortar were not used in Greek walls or foundations, hydraulic cement was known as early as the fifth century B.C. and was commonly used to line Athenian wells of that period. For foundations sited in marshy areas, Greek builders normally constructed a platform of rubble to support the foundations proper. Indeed, Plutarch describes such a foundation bed in his account of the construction of fortification walls through marshy territory (*Life of Kimon,* 13).

Systematic excavation of temple foundations can reveal not only the shape but also the procedures behind foundation construction. Excavation of the Hephaistion at Athens (ca. 450–445 B.C.), for example, demonstrates that instead of first building the cella and then adding the colonnade around it, as would seem to be the normal sequence of construction, the order was reversed. The foundations of the cella were clearly dug through the chips of stone left behind as the stylobate and its foundations were being built (figure 2.11). This surprising sequence of construction, which must have been somewhat awkward and inconvenient, undoubtedly had its motivation in precise visual alignment of cella walls with exterior columns (Coulton, 66–67).

The relatively small scale of Greek buildings, and the fact that their foundations were normally taken to bedrock, reduced the problem of foundation design to a relatively simple level. The requisites for imperial Roman public buildings proved to be an entirely different matter.

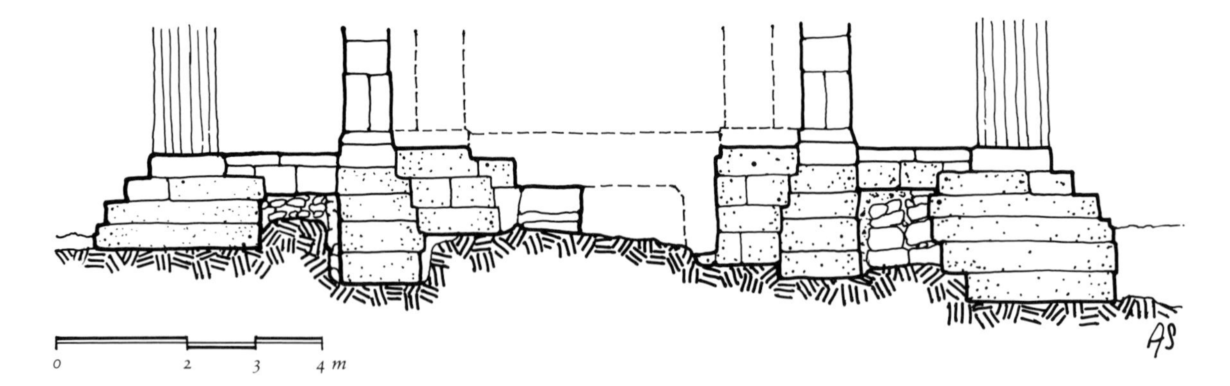

2.11 *Hephaistion, Athens, ca. 450–445 B.C.: section through excavation (after Coulton).*

Imperial Rome

Roman architecture, beginning in the late Republican period, set itself apart from Greek precedents by using new materials and new forms. Brick and concrete were employed in the creation of public buildings with spacious, vaulted interiors. For the underpinnings of these architectural experiments, Roman builders similarly introduced new technical solutions.

When building on a previously occupied site, the Romans did not necessarily demolish and remove all traces of preexisting buildings. Earlier buildings were often consolidated with rubble and incorporated into the foundations for the new construction. Standing walls from the Esquiline wing of the Domus Aurea, Nero's Golden House in Rome, constructed in A.D. 64–68, were reused in the foundations of the Baths of Trajan (dedicated in A.D. 109). Flat platforms, involving large earth-moving operations, were also used. The construction of Domitian's palace on the Palatine, the Domus Augustana, begun in A.D. 92, was preceded by leveling the site of the lower palace and terracing the upper level with concrete retaining walls to hold the displaced earth from the site. Domitian's engineers must have been highly skilled in earth moving, since they cut away the entire spur of earth linking the Capitoline and Quirinal hills. The scale of this enterprise can be appreciated by the fact that the top of Trajan's column (see chapter 5) marks the original height of the hill.

After site preparation, foundation trenches were normally dug, following Greek practice, to bedrock. In the case of the Temple of the Capitoline Jupiter in Rome (A.D. 75), this called for foundations five meters deep. For more modest buildings, especially in the northern empire, foundations might descend to only just below the frost line. Just as the lack of good building stone in Rome influenced the introduction of brick-faced concrete for walls (see chapter 3), Roman geological conditions also affected the design of foundations. Because of the ubiquitous presence of volcanic sand in the capital, foundation trenches could seldom be cut without interior wooden revetment, as illustrated in figure 2.12. The remains of these retaining revetments are visible in negative impressions on the concrete foundations of many Roman structures.

The most important Roman innovation in foundation construction seems to have been their massive concrete platforms. These are in some ways indebted to Greek practice, but were improved by

2.12 *Casa dei Dipinti, Ostia, early second century A.D. Revetment reconstruction (after Adam).*

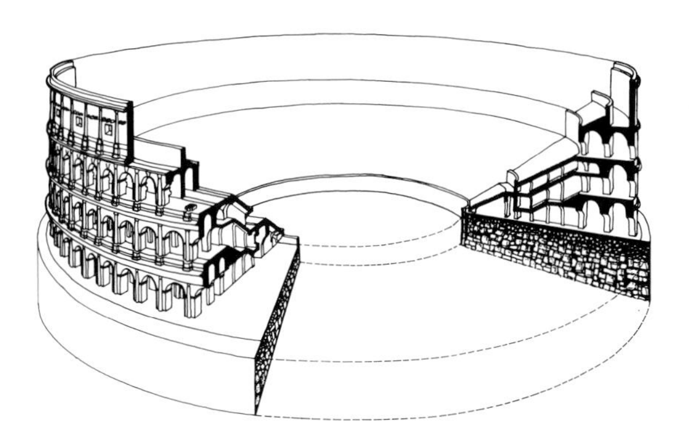

2.13 *Colosseum, Rome, A.D. 80: analytical drawing of the superstructure and foundations (after Lamprecht).*

the use of concrete. The hydraulic capabilities of Roman pozzolan cement allowed foundation platforms to be laid even under water, as for example at Ostia, the harbor town of Rome. Examples of pozzolan concrete footings are found under the Red Basilica in Pergamon and the Church of St. Polyeuktos in Istanbul (Mainstone, 176). The continuous platform foundation (today called a matt or raft footing), if sufficiently thick, represents a costly but appropriate solution to the problem of building on poor soil, in that differential settlement is effectively eliminated and bearing pressures are minimized by spreading building loads over the widest possible area. And, as already noted, when the weight of the foundation is great compared to that of the structure above, many of the problems resulting from soil settlement can be corrected *before* significant construction of the building superstructure.

The Colosseum in Rome, the largest of the Roman amphitheaters, stands on the site of the former lake of Nero's Golden Palace. Apart from the political advantages accruing to Vespasian for replacing a despised tyrant's "private" lake with a monument intended for public entertainment, the site provided a consistent subsoil for supporting the enormous stone oval (Sear, 135ff). A massive ring twelve meters deep, consisting of concrete and huge stone blocks (figure 2.13), supports the Colosseum (Sear, 71; Lamprecht, 157). Similarly, the Pantheon rests on a solid, continuous ring of concrete that is only 4.5 meters deep but over 7 meters wide (Adam, 115–117, 137–181, 199).

The Constantinian basilica of St. Peter's was sited on a flank of the Vatican hill over a cemetery that, according to tradition, was the location of the grave of St. Peter. The crossing of the basilica was placed close to ground level above the apostle's grave, and to maintain a level floor the nave was supported by fill more than 11 meters deep in the southeast corner and estimated to be more than 30,000 cubic meters in volume (Toynbee and Ward-Perkins, 197). The raised floor also allowed the builders to leave many of the graves undisturbed under the fill (an important issue because traditional Roman law prohibited disturbing graves). The walls of the small shrines built over the graves were left intact, and these, together with newly constructed masonry walls, served to subdivide the large volume of fill into smaller units. Continuous foundation walls, carried down through the old cemetery to undisturbed earth, supported the arcades and the outer walls of the church. The foundation walls were tapered only slightly toward their bases: measuring just over 2 meters in width at the foot of the nave arcade on the southern side, and about 2½ meters at the foot of the outside wall and aisle arcade. The builders could easily have provided a wider footprint on the soil, but they seem to have been unwilling to disturb additional tombs and/or their experience with earlier timber-roofed basilicas indicated that these structures (which were appreciably lighter than masonry-vaulted buildings) did not warrant wider foundations. Because of the proximity of natural springs in the Vatican hill, however, the resulting high water table reduced soil bearing capacity to such an extent that a 1505 report, indicating that foundation settlement had caused substantial movement of the walls, spawned the decision to tear down the Constantinian church.

Roman foundation walls had been built mainly of faced concrete, but in northern Europe, where dense silts and clays prevail, foundation trenches were cut to hold strip footings built of both

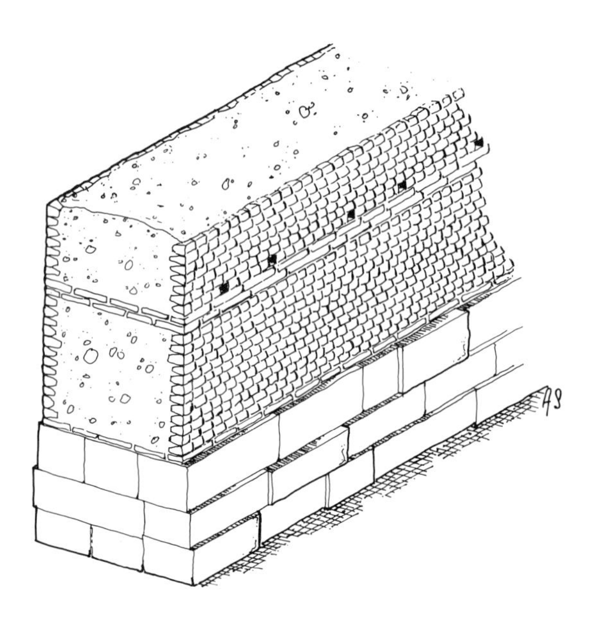

2.14 *Roman wall foundation at Beauvais, ca. fourth century (after Adam).*

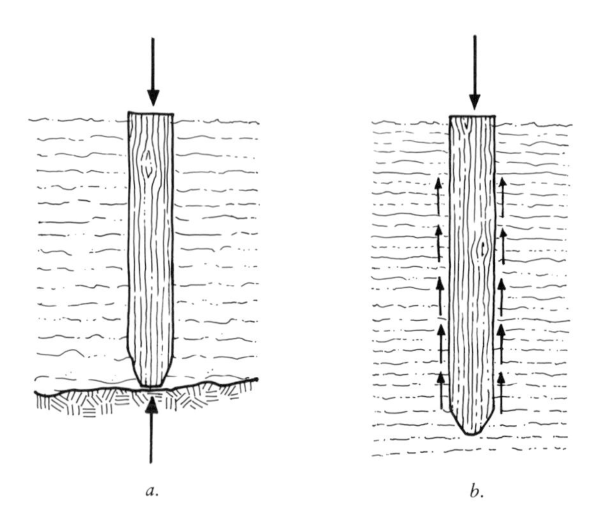

2.15 *Pile foundation types: (a) point-bearing, with primary vertical support from lower stratum; and (b) skin-friction, with vertical support from surrounding soil.*

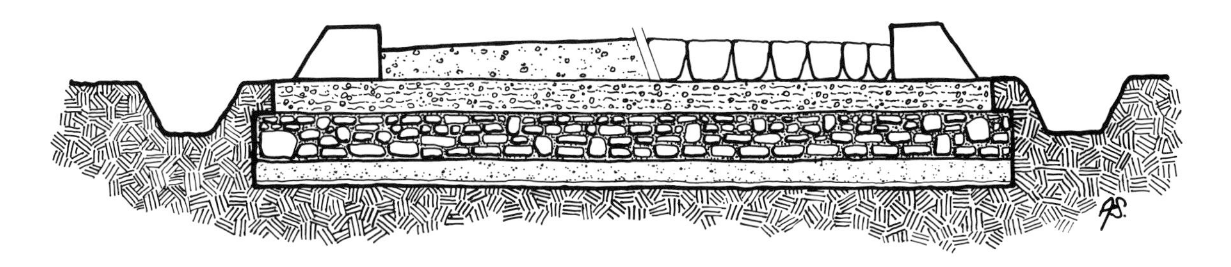

2.16 *Roman roads: typical cross-sections showing bottom sand layer and two types of upper layers: stone slabs (right) and gravel-concrete (left) for different expected wear (after Mainstone).*

stone and concrete. Characteristic of these regions are the massive rampart walls dating from the waning years of the empire. Most are constructed on wide, deep foundations built of large ashlar blocks from monuments sacrificed to the defensive campaigns (figure 2.14). Often, too, stone was selectively employed within concrete foundations where there were to be concentrated loads.

Hydraulic pozzolana was used for footings of bridges, as well as in the ubiquitous underground water courses and cisterns built by the Romans. Timber was also employed when foundations needed to descend below the water table. Vitruvius prescribes the use of **piles** (lightly charred tree trunks) for foundations built in marshy conditions (Vitruvius, V:12). This technology was commonly used for Roman fortification walls, especially in wetland sites. Vertical piles may be point bearing, that is, they translate building loads through soils of insufficient bearing capacity to a platform or soil of sufficient capacity beneath, or they may depend upon shear resistance between the soil and the outer skin of the pile (skin friction), as illustrated in figure 2.15. Although widely used in this way, the use of timber in foundations was not confined to vertical stakes. Layers of horizontal beams are interspersed between the foundations and the wall itself in a number of Roman fortifications. This feature, used in Pevensey and other Saxon shore forts in Britain, was also widespread on the continent. At the fort of Dax in France, where subsoil conditions were marshy, layers of timber beams were used under the walls; and at Strasbourg, the external towers were tied to the town wall by horizontal beams (Johnson, 32–33).

The towns and fortifications of the vast Roman Empire were linked by a remarkable system of roads, a few of which survive to the present day. The Roman roadbed represents a masterful foundation design usually comprising four superimposed layers as illustrated in figure 2.16. Sand in the bottommost layer "sealed" the road from moisture below to prevent frost heave, and the density and hardness of the uppermost layer, composed of stone slabs or of gravel-concrete, would have been selected according to expected traffic. Indeed, the introduction of concrete footings and the versatility with which public buildings, fortifications, and roads in a variety of soil conditions were built indicates the Roman development of a remarkably high level of foundation design.

Byzantine

Byzantine footings exhibit many of the advances made by the Romans, particularly in the area of hydraulic foundations. In drier zones of the Empire, stone **cisterns** lined with hydraulic mortar often were incorporated into the subterranean level of a building. Istanbul still preserves a number of examples, most built during the Constantinian to Justinianic periods (fourth through sixth centuries) to store water for a city susceptible to drought and siege (Mango, 123–129). Many of these are remarkably large, with the most famous being the Yerebatan Sarayī (in Turkish, the "Sunken Palace") or the "Basilica Cistern," measuring 140 by 70 meters and whose vaults were originally supported by 336 columns. The nearby Binbir Direk ("A Thousand and One Columns") cistern is smaller than the Basilica Cistern in plan (64 by 56 meters), but its (actual) 224 columns achieve greater height (15 meters) by being made up to two separate shafts joined at mid-height (figure 2.17).

2.17 *Binbir Direk cistern, Istanbul, ca. sixth century (Wulzinger).*

Until recently, our knowledge of the foundation systems beneath the vast churches of the early Byzantine era has been limited. If the foundations of Justinian's imperial church of Hagia Sophia in Constantinople extend to bedrock as is generally assumed, some of the credit for the long life of this great building must be assigned to the successful translation to ground of the enormous loading of its massive superstructure (Emerson and Van Nice; see also chapters 3 and 4). Limited excavations, however, have not yet confirmed this commonly held supposition. Recent, more extensive excavations of the smaller Haghia Sophia in Thessalonike (begun ca. 618), on the other hand, provide a well-documented example of Byzantine foundation practice (Theocharidou). A system of stepped foundation walls carry the superstructure loadings down to a level of clay 4.5 meters beneath present floor levels (figures 2.18 and 2.19). These walls, constructed in

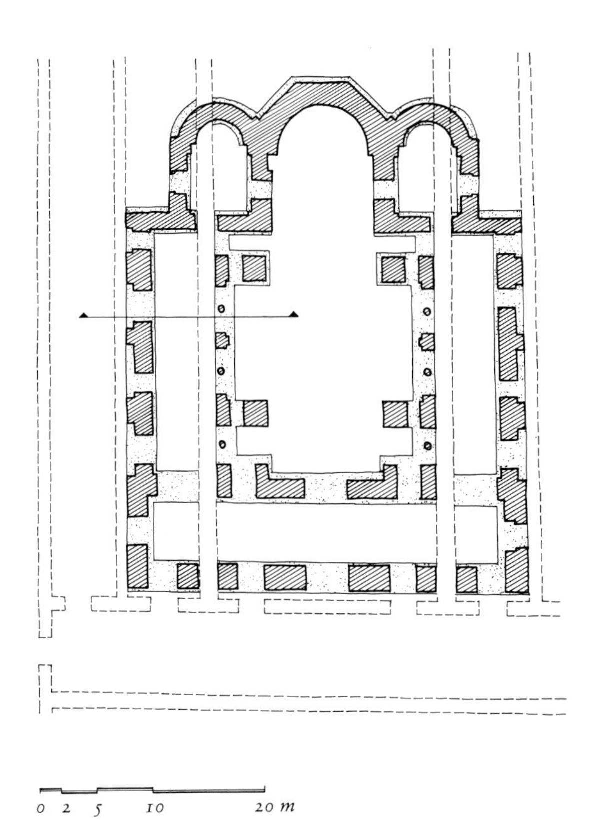

2.18 Haghia Sophia, Thessalonike, begun ca. 618: plan (after Theodcharidu).

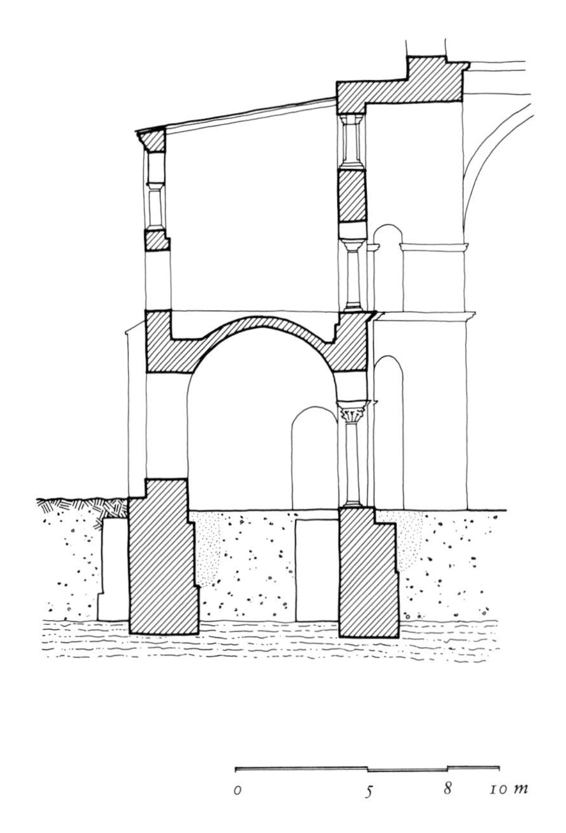

2.19 Haghia Sophia, Thessalonike: section (after Theodcharidu).

the seventh century, are built of limestone and hydraulic mortar and are laid parallel to the foundations for an earlier Christian basilica that had occupied the site. The later foundations extend deeper than those of the basilica, through a deposit of silty fill, and past the water table, to a layer of stiff clay. The seventh-century architect designed a single unit of continuous footings supporting the perimeter walls, and another under the piers and columns of the central unit. The two units were linked in two places, at the east and west. It is interesting to note that the architect carefully avoided placing any part of the new church directly on the old foundations, even to the point of creating oddly syncopated arcades. It is likely that the seventh-century architect recognized the potential danger of foundations resting on weaker soils located at the level of the water table, and therefore did not trust the footings of the preceding basilica.

The provision of separate foundations in a single building campaign is also found in the large twelfth-century church now known as Fatih Camii at Enez (figure 2.20). Since the narthex is not bonded to the main body of the church, some historians have concluded that it must have been a later addition. Close relationships of constructional and ornamental elements suggested to Robert Osterhout, however, that lack of bonding is due to a difference in the mass of the two building units rather than different building campaigns (Osterhout 1983, 273). The

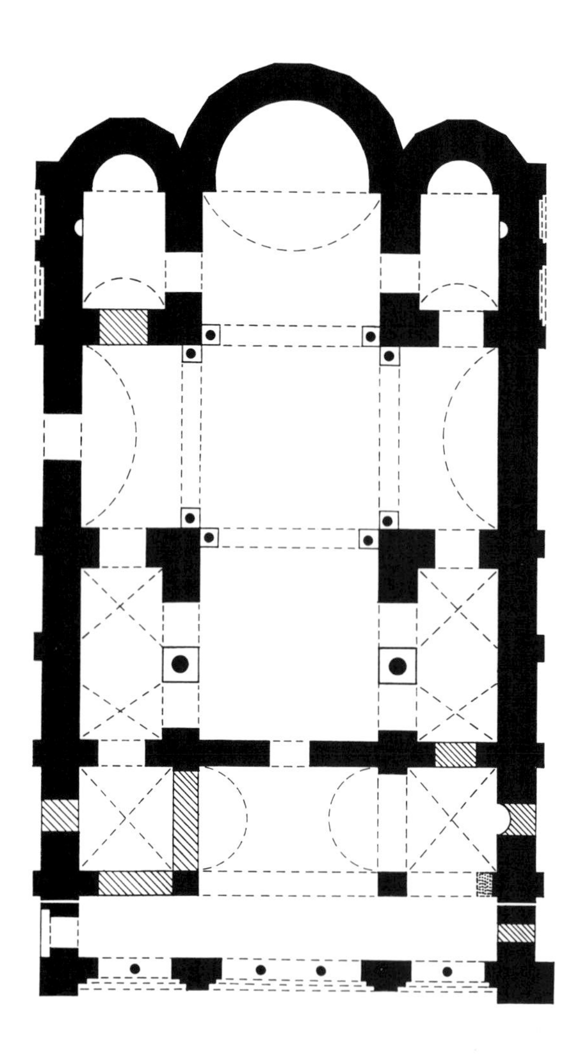

2.20 *Fatih Camii, Enez, ca. twelfth century: plan. Note discontinuity between the church proper and its narthex (Ousterhout).*

main church, domed and vaulted in stone, is much heavier than the light timber-roofed narthex, and would therefore have more prone to settle. The twelfth-century builders seem to have wisely designed foundations of different depths and/or profiles for each building unit, and left the two units unbonded to allow for differential settlement. Increasingly, excavations have revealed the presence of earlier foundations at a number of Byzantine sites. It remains to be seen whether Byzantine builders reused these preexisting foundations, as they did at Kariye Camii (Ousterhout 1987, 13ff), or if they avoided them for technical reasons, as did the architect at Haghia Sophia in Thessalonike.

Early Medieval

Those few architectural studies that have discussed early medieval foundations have generally asserted that they are shallow, roughly built, and "inefficient" (e.g., Kerisel, 41, 49). These suppositions treat only the formal, "architectural" design of foundation walls and not the underlying soils or the relationship of foundations to those soils, disregarding the larger context in which a foundation serves. Moreover, this negative view of early medieval foundations ignores examples like the Carolingian foundations of the basilican cathedral of Cologne, which were so effective that they were partially reused by the Gothic master in the massive thirteenth-century rebuilding of the cathedral (Doppelfeld, 55ff).

Excavations at the Anglo-Saxon church of Brixworth provide one of the clearest examples of early medieval foundation practice. Those excavations clarified the original plan of the church, later altered into an open basilican plan: a nave with five

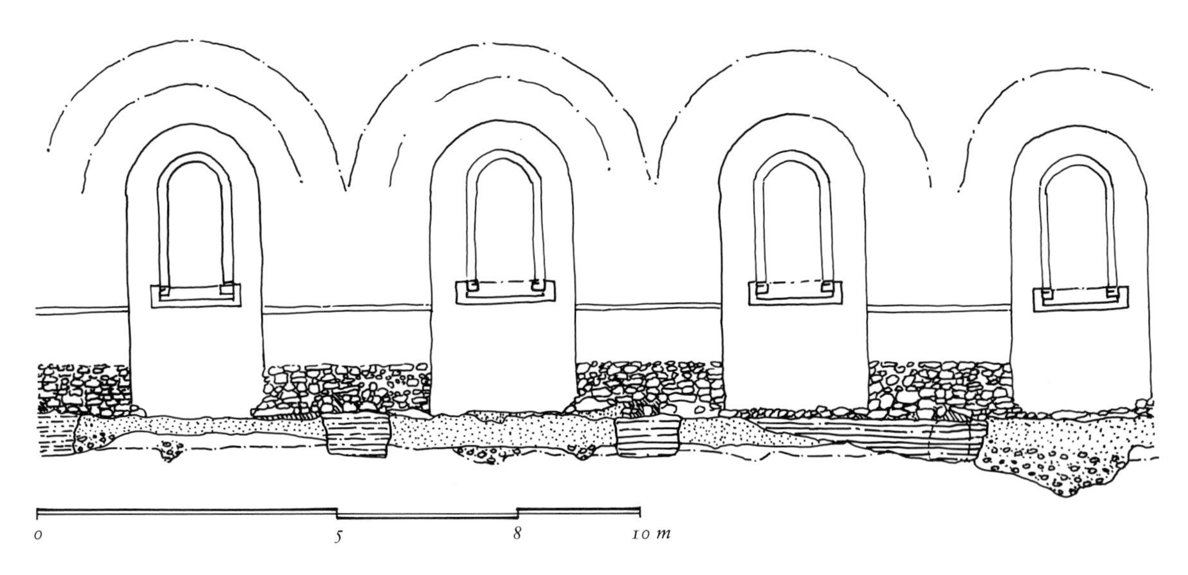

2.21 *Brixworth abbey, ca. ninth century: longitudinal section taken through foundations (after Audouy et al.).*

porticus projections to the north (and probably also to the south). The foundations were constructed of small-scale ironstone rubble blocks from 60 cm to 1.6 m in depth, as shown in figure 2.21. This variation seems to have occurred where the builders questioned the stability of the underlying soil, such as those areas where they encountered loosely packed ditches from earlier construction on the site. Terracing was performed to bring a sloped site to level, procedures also linked to foundation construction. Stone and mortar analysis confirm that the foundations were all laid during a single construction period, although butt-joints and thin layers of silt attest to short periods of delay within the process (Audouy et al.).

The foundations at Brixworth, though relatively shallow and roughly built, effectively supported the loads of the church walls. The almost exclusive use of a hard ironstone signaled a concern for suitable masonry. Although denser stone was not a structural necessity, it was commonly used throughout the medieval period, being employed, for example, in the foundations of Amiens Cathedral. Yet from a technological perspective, the early medieval period in the West seems to have contributed no new developments to foundation design or construction. This continuity may well be the result of the excellence of Roman building practice as well as the relatively modest scale of most early medieval monuments. By virtue of their reduced scale, these buildings could perform adequately with simple foundations.

Romanesque

As with early medieval construction, the generally held assumption that Romanesque foundations are shallower and less well built than those of the Gothic rests less on real knowledge than on the unspoken assumption of "progress" in architectural technology. Recent archaeology, however, has revealed examples of ashlar foundation walls and of a stepped profile for those walls in Romanesque buildings. Excavation evidence also makes it clear that Romanesque builders were well aware of the necessary relationship of their foundations to surrounding soils.

The Romanesque third abbey church at Cluny, one of the largest and tallest buildings of the Middle Ages, was built westward, from the east, during the late eleventh and early twelfth centuries. Its surviving transept vault soars almost 30 meters above original floor level. With vaults spanning 10.5 meters across a nave equal in height to the transept, the construction of stable foundations for the giant church must have been a high priority and, thanks to soundings undertaken by K. J. Conant in the late 1920s, we know something of these (Conant 1929, 1931; Kerisel, 49–50). The utility of this information is limited, however, because Conant undertook only small soundings in certain areas. On the other hand, Conant, unlike many excavators of his generation, took his soundings to the base of the foundations, describing briefly the observed soil conditions.

The foundations of Cluny, where revealed, were built of rough-dressed ashlar, which faced a rubble core. At the eastern end of the church (in the axial chapel) and in the little transept, the builders descended to a layer of gravel that Conant identified as marking the water table. The base of the foundation at the axial chapel lay about 4 meters below modern ground levels, but only slightly less than 2 meters below the medieval ground level. This gravel layer, which lay deeper (relative to the ground slope) toward the west of the church, was probably the best geologic substrate for large-scale building in the river valley where Cluny is sited. Toward the western end of the nave, we know only that the foundations were shallower, and therefore they probably did not reach the gravel layer.

One may question the contribution of geologic conditions and the foundation system of Cluny III to the collapse of the nave or transept vaults in 1125 (although another possible source for the Cluny collapse is offered in chapter 3). Sharp changes in the groundwater level might have contributed to the destruction if the builders did not excavate to the deeper, gravel layer toward the west. Documented shortages of time and funds during the early twelfth century could well have contributed to the shallow depth of the western foundations. But such questions await the results of excavation now in progress below the western portion of the site with careful attention being paid to both stratigraphy and geology.

An interesting contrast to the foundations of Cluny III was observed by the writers while inspecting excavations at the late Romanesque monastery church of St. Lucien in Beauvais. Now destroyed, St. Lucien was a fairly large church, some 80 meters long, and sited on a sharp slope above the medieval city. Recent excavations have revealed that while the church had enormous foundations for its western towers, at the east end there were no foundations at all. In effect, the western foundations created a platform against the slope of the hill. The western towers on the downhill side of the site required massive footings to rise to the interior floor level. At the eastern end of the church, the curving walls of the ambulatory could be set directly onto bedrock which, at that point, rose to ground level. Although one cannot be entirely certain because of the limited areas excavated, it can be assumed that the foundations descended to bedrock throughout, creating a massive terrace (triangular in profile) upon

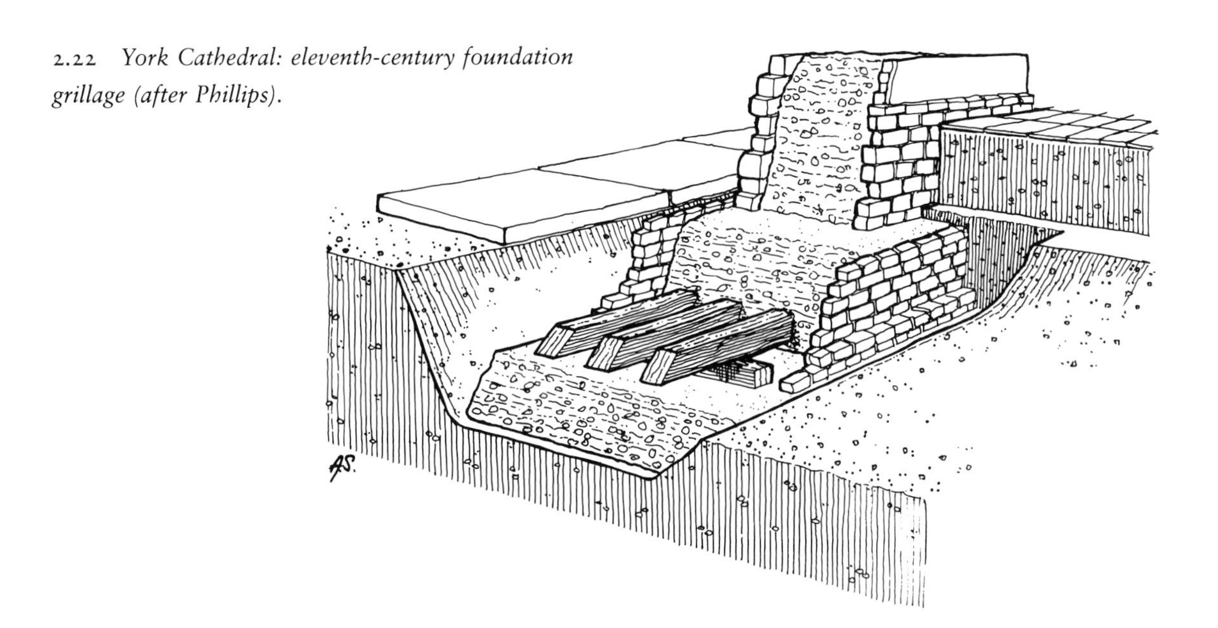

2.22 *York Cathedral: eleventh-century foundation grillage (after Phillips).*

which the church could be built. The foundations of St. Lucien thus represent another instance in which the builders, quite sensibly, seem to have designed the foundations for a particular local site geology.

Horizontal **timbers** were also used in a variety of Romanesque foundations including the substructures of two large Anglo-Norman cathedrals. A system of substantial beech logs was imbedded in mortared rubble to produce a continuous strip footing under the east wall of Winchester Cathedral where the supporting soil was a saturated peat bed. This timber-reinforced footing functioned effectively until the early twentieth century when a falling water table allowed the peat bed to consolidate and exposed the timbers to the air (Bussby 1979, 1987).

Excavations at the contemporary York Minster in the 1960s revealed a remarkably sophisticated use of timber chaining. The eleventh-century foundations included a system of horizontal timbers set into mortared rubble to produce an interlocking reinforcement grid within the substructure (figure 2.22). This system gave the footings bending resistance, which helped to spread the weight of the building over a greater area.

The foundations at York descended to a layer of stiff clay. This three-meter-thick deposit, $4\frac{1}{3}$ meters under floor level, was identified by modern excavators as a "leveling deposit" laid down by the Roman builders of the *Principia*, which had previously occupied the site. The excavators speculate that the Anglo-Norman builders, who followed it consistently despite deviations in its level, may have mistaken this relatively deep and clean soil layer as natural clay in situ. The Norman builders would seem to have made a careful search for an acceptable load-bearing stratum before foundation work began, and probably monitored this chosen layer during the first weeks of work upon the substructure. It is interesting also to observe that the thirteenth-century builders at York reused the eleventh-century foundation system. They did so with good reason. Wherever they were forced by expansion to install their own foundations, they built neatly dressed ashlar walls. These Gothic foundations, however, did not descend to the safe level of stiff clay used by their Romanesque predecessors—which led, over time, to the critical problems of building settlement that necessitated massive intervention in this century (Phillips, 175).

In at least some instances, Romanesque foundations descended as deeply, and were better built, than Gothic foundations; though the limited evidence we have so far for Romanesque foundations probably should not be generalized. During the eleventh and much of the twelfth centuries, building was organized regionally. Evidence from northern France, an area rich in limestone deposits near the surface, may not be representative of France as a whole, much less of Europe. We also lack evidence about foundation construction for large-scale Romanesque buildings outside this region, such as St. Sernin in Toulouse. Much of what is characterized as "Gothic" foundation construction may well have existed beneath similar large-scale Romanesque churches. Certainly the timber-reinforced continuous footings used by Norman builders for foundations built on stiff clay are as advanced in concept and in execution as the continuous mat grid that would be developed by Gothic builders at Amiens 150 years later.

Despite numerous indications of successful consideration of the variables of scale, soil strength, and materials by ancient and medieval designers, notable failures such as those of the foundations of the Campanile of Pisa Cathedral (Leaning Tower; figure 2.6) demonstrate that knowledge of soil behavior was (and remains) far from an exact science. This structure has suffered both from excessive settlement and tilting due to shear distortion, which has progressed over the last 700 years to a point where collapse now seems imminent.

Construction of the tower began in 1174, but was suspended for nearly a century when the superstructure reached three and a half stories, apparently because of substantial vertical settlement (accompanied by a modest tilt to the northwest). Work resumed in 1272 and as the tower rose, it tilted successively toward the north, the east, and then when the tower reached the sixth story, it took an alarming tilt to the south. Construction was halted again in 1278 at the level of the seventh story, which was built as a wedge to compensate for the accumulated tilt. Almost 100 years lapsed until the final, eighth story—also wedge-shaped to accommodate the tilt by that date—was added in 1370.

From numerous soil samples taken over the last fifty years to determine the underlying soil properties, it is now possible to theoretically calculate the Pisa settlements. Indeed, the theoretical value agrees almost exactly with the measured average value of 1.8 meters. The tower foundation acts as a single footing with an average contact pressure of 51 tons/meter^2 and is rotating about a point near the center of its base. As the settlements have occurred, due primarily to water being squeezed from the upper clay layer (a phenomenon described in the introduction), the strength and stiffness of this relatively weak and soft clay has gradually increased, keeping pace with the added shear stresses along the slip surface caused by the greater offset of the tower from increased rotation. The tower now tilts at an angle of 5½ degrees, and is currently increasing its tilt at a rate of roughly 1.1 minutes per decade (Mitchell, et al.; Kerisel, 41, 42). How long this precarious balance can continue is unknown. As a rule, clays eventually lose strength when the shear distortion of the failure surface becomes very large. Indeed, the prognosis for the tower is not good unless remedial measures are initiated soon.

The designer of the Pisa tower in the twelfth century cannot easily be faulted since the problematic soft, weak clay layer is hidden below the 10 meters of sand upon which the tower foundations rest. There were no other structures in Pisa requiring anywhere as much bearing capacity, so no other foundation trenches would have gone deep enough to expose the hidden soil problem. Until the tilting instability began, the large uniform settlement, which was certainly unanticipated and already over 1 meter by 1270, was only an inconvenience requiring steps to be built. After the initial rate of tower tilt slowed, one can imagine that public pressure favored completion rather than destruction. As a result, the Tower at Pisa has stood for 600 years as a monument to the importance of site exploration and proper foundation design.

Gothic

Like the remarkably thin skeletal walls of Gothic architecture aboveground, Gothic foundation design has been similarly credited with innovation (Kerisel, 49–51). Stepped interlocking grids penetrating to great depth have been seen to be the underground counterpart to increasingly taller and thinner buildings. But rather than a formal development parallel to aboveground styles, Gothic foundations reflect the same pragmatic concerns and knowledge combined with appreciation of specific soil properties that we have seen in previous eras.

Nineteenth-century excavations of the foundations of Amiens Cathedral by Viollet-le-Duc and others have long served as one of the very few well-documented examples of Gothic foundation construction and have generally been regarded as typical of a Gothic cathedral (figure 2.23). Yet the extent of the area excavated at Amiens is unclear. At a minimum, one bay in the choir side-aisle and an area between the buttresses must have been opened, but a much larger area in the choir may have been exposed as well. In any case, it is certain that the foundations of the entire building have never been explored.

While the interlocking grid in the foundation walls at Amiens is similar to that used in the Greek temple of Apollo at Delphi (figure 2.9), it is more closely related to the Romanesque timber format at York even though the Amiens foundations descend much deeper than those at York. This increased depth is due not only to the greater height and weight of the immense superstructure, but also to the lack of a stable soil layer close to the surface, as can be established by comparing the bearing pressure and its distribution with the corresponding stiffness and bearing capacity of the supporting clay soil based on information from the recorded excavations (see figure 2.24).

Several observations are suggested from our analysis of the Amiens footings. First, the structure is remarkable for its efficient design: calculations based on conservative assumptions indicate that the supporting clay is stressed to about 65 percent of its allowable bearing capacity. Second, the bearing

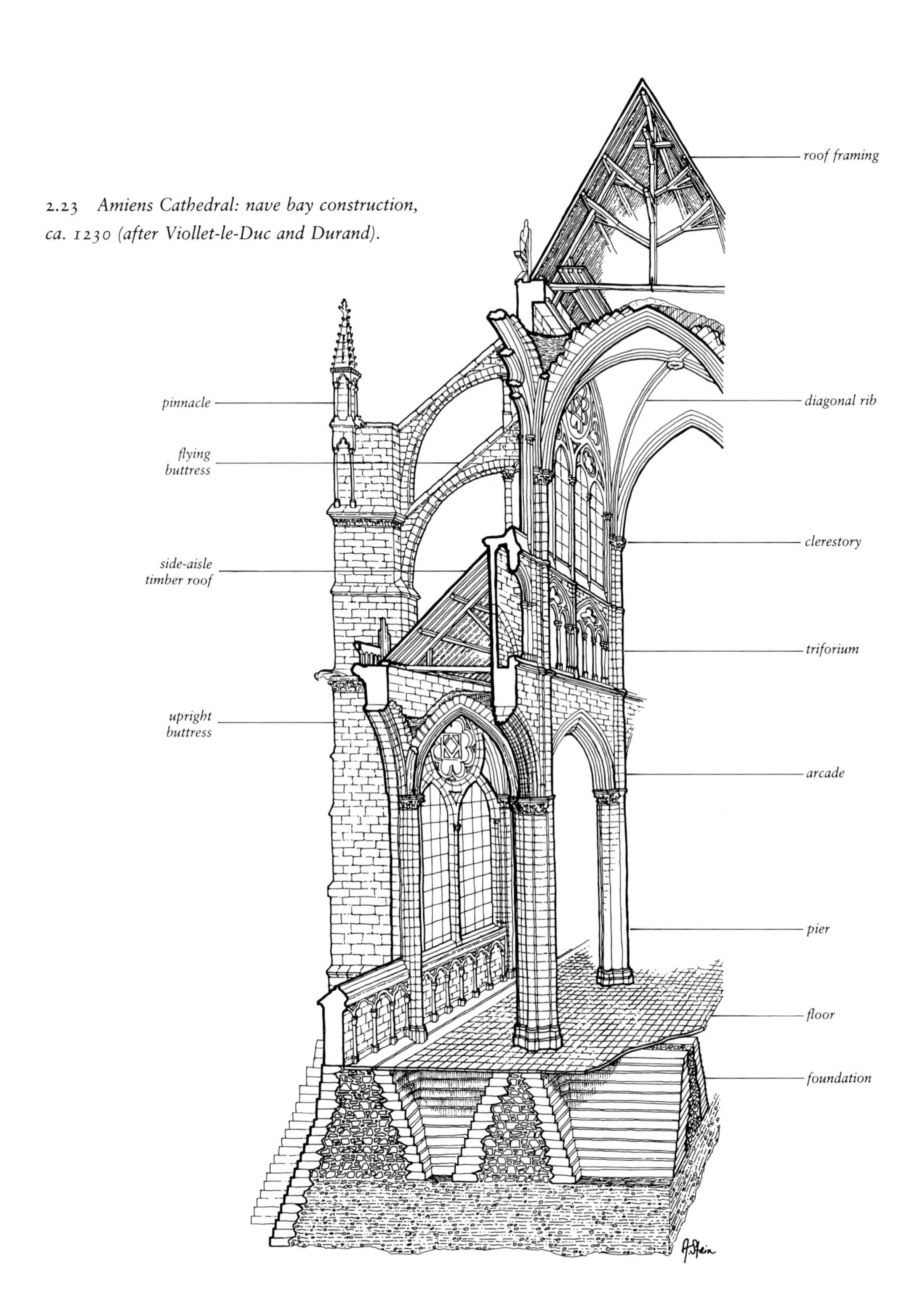

2.23 *Amiens Cathedral: nave bay construction, ca. 1230 (after Viollet-le-Duc and Durand).*

2.24 Amiens Cathedral: foundation analysis. *Since all the bay loadings are directed to individual piers and buttresses, these can be analyzed and generalized to the building as a whole. The distribution of the weight of the superstructure to individual piers and buttresses as well as the footing weights are derived from the type of model study described in chapter 3. The bearing pressure under the footings is then determined by dividing the weight of the footing and the superstructure loading by the area of the respective footing bases. For the piers, this was found to be 23.7 tons/m^2, and for the buttresses, 25.6 tons/m^2 (Bonde, Maines, and Mark).*

Viollet-le-Duc's excavation astride the buttresses revealed that the foundations rested on a "stiff" (hard) clay rather than the bedrock one might have expected for a building of this size. No test data is available for soil from the site, but the modern prescribed, allowable bearing value for stiff clays is given as 40 tons/m^2 (table 2.1), indicating safe design.

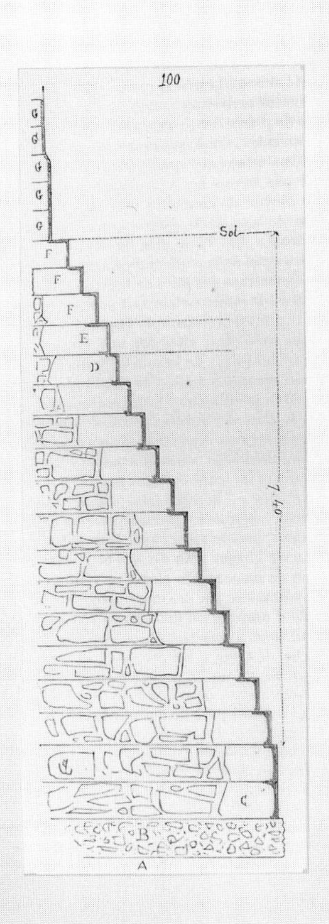

2.25 *Soissons Cathedral: nave foundations, ca. 1200.*

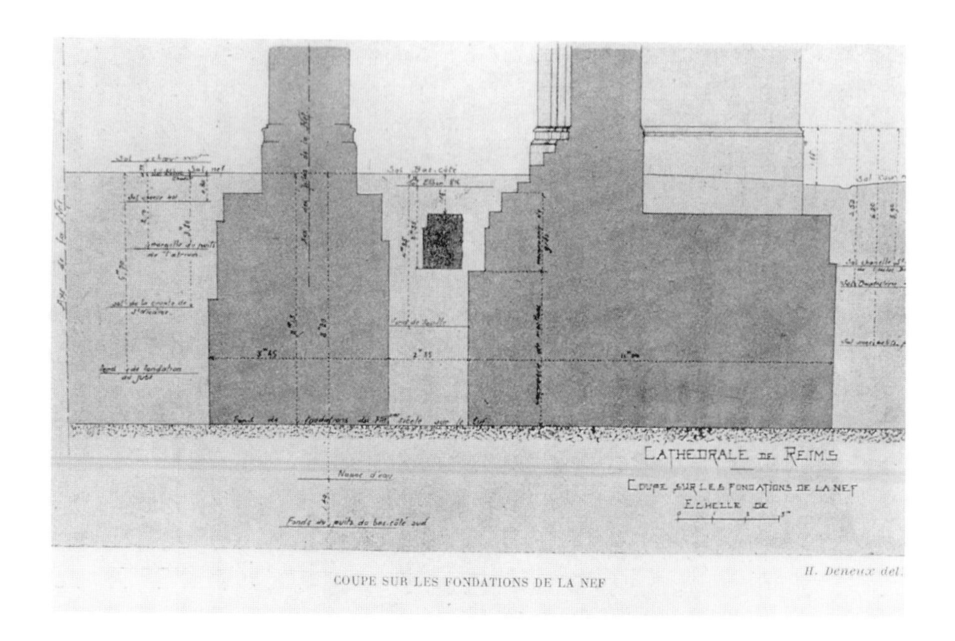

2.26 *Reims Cathedral, begun 1210: foundations below the nave (Deneux).*

stresses beneath the choir pier and beneath the pier buttress are remarkably close: 23.7 and 25.6 tons/meter^2 respectively. The similarity of these stresses all but eliminates the risk of differential settlement. Third, the foundations themselves are of such enormous size and weight that their construction alone would have revealed acute problems of instability or differential settlement before work on the superstructure began. It would seem that the builders at Amiens had a good understanding of the advantage of distributing the load of a tall building over a wide area. In constructing a broad, mortared, latticed raft, the builders seem to have quite deliberately distributed the building load. This decision would seem to have been particularly wise since the cathedral was built on a clay stratum adjacent to the river Somme, which could well have meant a high water table. The stepped foundation walls at Amiens were the only viable medieval solution for stone footings that needed to descend nearly eight meters. Narrow, unstepped walls would not have spread the load widely enough over the clay soil. Thus the stepped foundations at Amiens do not represent a stylistic solution to foundation design, but rather a logical, technological response to a particular site geology. Furthermore, the development of the interlocking grid at Amiens may have evolved out of a desire to stabilize large foundation walls that would otherwise have been supported laterally in the interior only by backfill.

In contrast to Amiens, evidence for portions of the foundations of the cathedrals of Soissons (figure 2.25), Bourges, and Reims (figure 2.26) suggest that longitudinal walls used with only occasional cross elements (or with cross-walls entirely absent) were more typical of large-scale Gothic construction than the interlocking grid. These longitudinal walls usually step outward two or three courses before they descend vertically without interruption. The cross-members appear at critical axes, like the crossing bay and behind the western towers, but not at every bay along the nave or choir. Deneux's excavations at Reims cathedral in 1925, for example, revealed continuous wall foundations that descend eight meters to rest on natural chalk (Deneux, 9–10). Cross-walls are found only where the major vessels intersect.

2.27 *Saint-Etienne, Beauvais: early twelfth-century foundations below the choir piers.*

The results of excavations at many Gothic churches such as Saint-Jean-des-Vignes in Soissons, St. Etienne at Beauvais (figure 2.27), and the cathedrals of Chartres, Cologne, and Orléans show that they relied less upon newly built foundation systems than on those reused from an earlier structure on the same site. At each of these sites, where alignments with previous construction permitted, the Gothic builders set piers and walls of their new churches directly atop older foundations without making any attempt to shore up or deepen them. In so doing, the builders seem to have ignored any potential problems of differential settlement that reuse might have entailed. Evidently they believed that the old foundations were stable enough to support the new, usually larger building. In this they were generally correct, because the weight of the earlier structure would have compacted the soil layers below, making them both stiffer and stronger. Moreover, it is probable that the protracted periods of construction at these and other Gothic sites would have revealed differential settlement before the buildings attained their full height. In such cases, the builders might have had opportunity to strengthen the foundations before continuing construction.

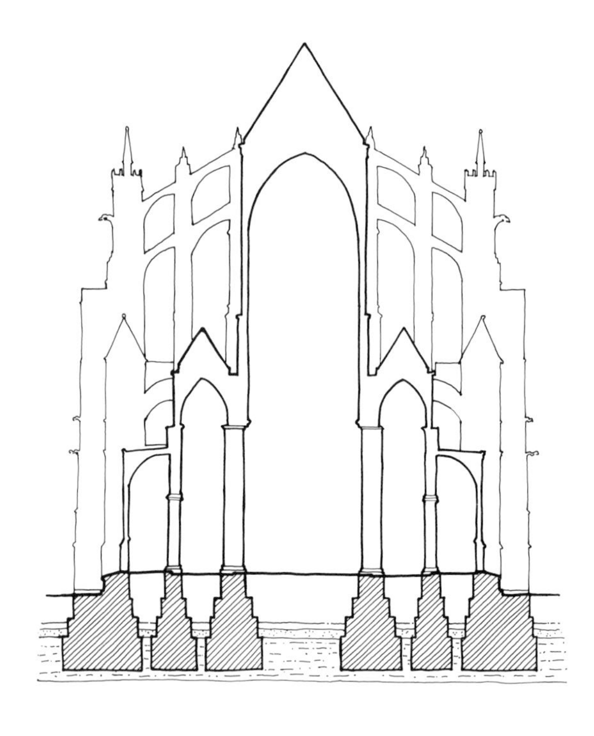

2.28 *Beauvais Cathedral, choir, begun 1225: diagramatic section indicating local soils and footings (after Chami).*

The foundations excavated beneath the eastern claustral range at Saint-Jean-des-Vignes, Soissons offer further testimony to a pragmatic knowledge of site geology on the part of Gothic builders. The abbey is sited on a hill whose contours are defined by very stiff, dense, layered clays. Its eastern range, destroyed during the 1830s, was a long, rectangular, two-storied building sited at right angles to the slope of the hill. Excavation within the chapter room demonstrates that Gothic builders on the site were aware of the natural geology and that they manipulated it to best advantage (figure 2.3). In constructing the eastern claustral range, as elsewhere on the site, the builders seem to have identified a single geologic layer of dense clay as strong and stable. They followed it across the entire width of the range despite its varying depths from ground and floor levels. The two inner piers of the chapter room carried identical loads, yet the western pier footing descends less than a meter while its neighbor to the east descends more than a meter and a half. This difference in depth of foundations does not seem to be caused by the slope of the hill, which in any case was effectively eliminated by terracing. The foundation walls are constructed of rough reused blocks, many probably taken from the Romanesque abbey. These rubble-built foundations were unmortared, possibly to permit drainage through the clay soil. Pier footings look, in section, remarkably similar to those at Delphi (figure 2.9) or at Brixworth. But rather than descend differentially to bedrock (as at Delphi) or through soft backfilled pits (as in the foundations of the church at Brixworth), the builders of the chapter room at Saint-Jean-des-Vignes were consciously seeking a known soil layer.

One final example of the pragmatic understanding of soil bearing capacities in relation to building size can be seen at Beauvais, the tallest of all Gothic cathedrals. Excavations there, being carried out under the direction of Emile Chami, reveal that the thirteenth-century builders of the choir dug through both the stone raft of the church of Notre Dame de la Basse-Oeuvre and a deeper Gallo-Roman foundation, as well as a weak layer of peat to reach the level of natural chalk (figure 2.28). We know from documentary evidence that later, in 1499, trial pits were dug by a new architect of the works, Martin de Chambiges, presumably to locate this same safe level of natural chalk in order to extend the unfinished building further to the west. At Beauvais, knowing the intended height of their church during both periods of construction, the builders judged that neither the older foundations on the site nor the layer of peat beneath them could adequately support their new cathedral. Their search for that chalk substrate and the resulting nearly 12 meter depth of the foun-

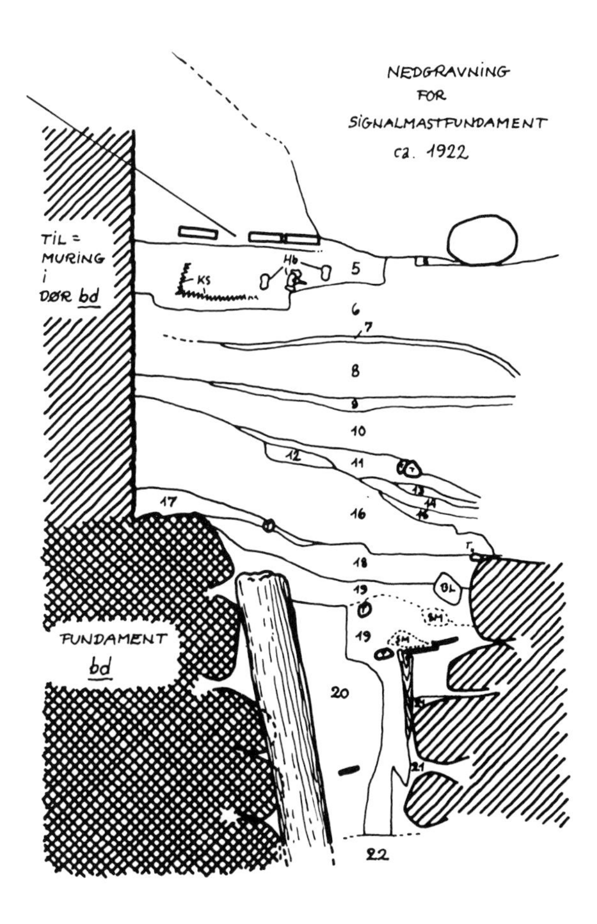

2.29 *Franciscan Monastery, Svendborg: thirteenth-century timber-pile revetment (Jansen).*

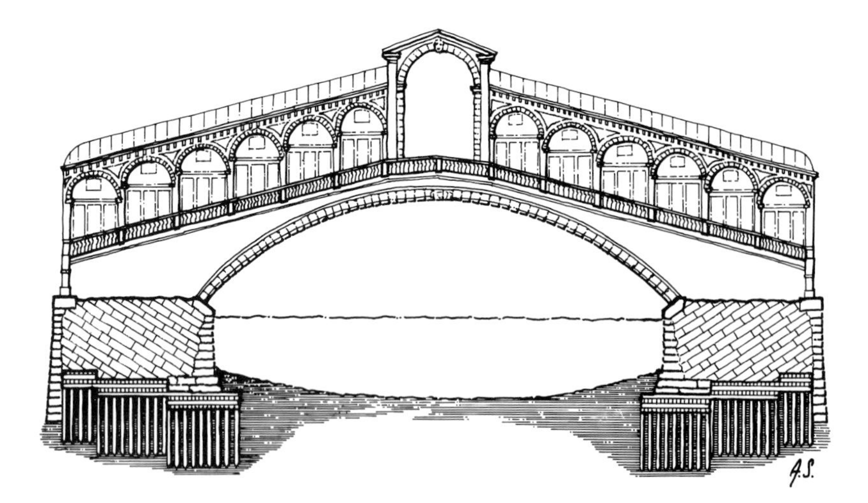

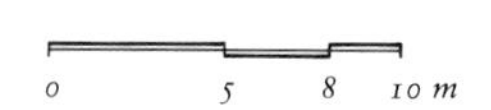

2.30 *Rialto Bridge, Venice, ca. 1340: section indicating installation of extensive timber piling.*

dations at Beauvais should be seen as a reflection of that considered judgment.

Like the Romanesque foundations beneath the cathedrals at York and at Winchester, Gothic foundations also employed timber, especially on marshy sites. The thirteenth-century Franciscan monastery in Svendborg in Denmark used vertical piles next to the foundations (figure 2.29) apparently as revetment in waterlogged construction conditions in addition to horizontal timbers underneath the footings. Vertical piles were used in numerous medieval bridges (figure 2.30) as well as for structures built in waterlogged sites. These include the thirteenth-century Old City at Stockholm and much of the city of Venice, including its cathedral of St. Mark's as well as most of the medieval and Renaissance houses whose support depended upon a system of arcades resting on piles.

The excavated evidence for Gothic foundation design presents a wide range of solutions for providing a solid, stable substructure for a massive stone building. Current evidence suggests that the design of the foundations under Amiens Cathedral is the exception rather than the rule. The consistent reuse of earlier foundations (and other building materials as well) suggests the pragmatism and economy of Gothic builders. Timber piles employed on waterlogged sites demonstrate a response to local geological conditions as opposed to a simple formal development of foundation walls.

Renaissance

Treatises of the Renaissance are more comprehensive in their recording of foundation practice than medieval records and tend to confirm what we know archaeologically from earlier centuries. Leon Battista Alberti (ca. 1404–1472) describes being present when Roman ruins were being excavated, and architects from Brunelleschi to Palladio (see chapter 3) actively studied early building sites from both a design and construction viewpoint. The observation of ancient foundations was particularly helpful for Renaissance architects, because large-scale buildings naturally posed the greatest difficulties. And, as discussed in chapter 6, the proliferation of building treatises disseminated such knowledge throughout the literate architectural community.

Foundation design seems to have been governed by simple proportional rules. Palladio states that foundations should be twice as thick as the walls they support, which proportion should be altered with regard to the "quality of the ground" and the scale of the building (Palladio, I. 8. 1). Likewise, Alberti recommends piles to be no less than one eighth the planned height of the wall to provide proper support. Walls should be placed over the center of the footing, and the foundation excavation should be absolutely level to avoid any slippage or movement (Alberti, III. 3). Serlio alludes to the widespread adoption of such rules, indicating that "I need not show [footing details] here, because every man knows it" (Serlio, IV. 14. fol 1).

Alberti also discusses types of soil. He advises, for example, that foundation trenches be dug down to "solid ground," avoiding underground springs. He even recommends avoiding soil with rounded stones (which would indicate the presence of water or an alluvial deposit) in favor of soils with hard, sharp stones. Several wells or pits should be dug to determine the types of soil strata present under the surface. When a "hard layer" is reached, it can be tested by placing a bowl of water on the surface, and then rolling or dropping a heavy object on the ground. If the water remains undisturbed, the soil is hard enough. Alberti, however, reserved as the final authority "local residents or nearby architects: through their acquaintance with existing buildings or their daily experience in constructing new ones, they will have acquired a ready understanding of the quality of the local soil" (Alberti, III. 1. 2). The observation of previous building technique also taught Alberti that foundation stones need not be of the highest quality. Since they are not exposed to weathering like aboveground courses and are confined by the surrounding soil, any stone will be adequate "so long as you do not fill the foundations with rubbish and material that might be perishable" (Alberti, III. 6). Caution, too, is given against reusing existing foundations unless careful evaluation is made of their solidity.

Although Renaissance architectural treatises demonstrate a good understanding of foundation behavior, the ability of architects to judge the strength of soils and foundations was only qualitative. Standard-sized buildings on known soils would thus be correctly designed, but changes in scale or site were likely to cause difficulty. Evidence of this is found in the construction of the new St. Peter's Basilica in Rome. The scale of Bramante's design was far greater than any recent Italian building, with the exception of Florence Cathedral. Because he did not have sufficient experience with buildings of this scale, Bramante's original foundations had to be deepened and enlarged first by Sangallo and later by Michelangelo. Yet despite this remedial work, differential settlement of the southern piers later caused cracking in the arches they support, requiring repairs during construction of the dome and again in the eighteenth century (see chapter 3).

Unlike St. Peter's, most churches of the Renaissance were relatively small, being closer in scale to early medieval buildings than to the larger and heavier Gothic cathedrals with their deep foundations. Excavation beneath the numerous civic towers of Italian Renaissance cities would provide interesting insights into foundation design of heavier buildings in the Renaissance period. Yet, despite the lack of physical information about Renaissance foundations and adjoining soils, the Renaissance building treatises of Serlio, Palladio, and Alberti attest to a significant degree of continuity with older foundation practice with their close attention to local soil conditions tempering general proportional rules.

No neat evolutionary development of foundation design can be traced through the two thousand years of premodern architecture under scrutiny here. In technique and form, foundations were as varied as the buildings they supported, and the soils into which they were laid. Builders from the Greek to the Renaissance periods found a number of effective solutions to translate loads from wall to supporting soils, and both continuities as well as discontinuities in foundation practice are evident. While the concrete that was a staple of Roman construction disappeared from European medieval architecture, interlocking grid foundations of stone are found at both the Temple of Apollo at Delphi and the cathedral of Amiens. Foundation walls might be rough and unmortared, as in most of the Greek, early medieval, or Romanesque foundations, or they might be constructed of carefully cut ashlar blocks mortared into subterranean "walls," as at Amiens. The reuse of foundations from predecessor buildings on the site is another prevalent feature found in Roman, Byzantine, and especially Gothic construction. Indeed, the constraints imposed by reused foundations may have influenced the plan, dimensions, and design of more buildings than we have realized. One constant of foundation design, especially for the large-scale structures that are the particular focus of this book, was the reliance on soil strata identified by historic builders as effective. We have seen the use of "telltale" layers at the Byzantine church of Haghia Sophia in Thessalonike, the Romanesque abbey church of Cluny III, the Gothic cathedral of Amiens, and the thirteenth-century chapter room at Saint-Jean-des-Vignes.

Any survey of foundation practice during the era before the Scientific Revolution also reveals the incomplete nature of our evidence. The need for more foundation and soil information is acute. This information can only come from excavation, with full recording of soil conditions adjacent to excavated foundations. Analysis of historic buildings will not be complete until such a corpus of comparative material has been compiled. The history of foundations we have traced above may well require drastic revision when a larger group of foundations is known. This chapter is offered as a preliminary outline and as an impetus to further study of foundations which, we believe, is essential to a more complete understanding of the design, the sequence of construction, and the structural behavior of historic buildings. This general discussion of foundations has introduced a constellation of themes—experience as the guide to building practice, the central role of geometry as a tool of design and analysis, the importance of scale, and the fecundity of innovation—issues that will resonate again throughout the following chapters on walls, vaults, domes, and timber roofs. Rather than inchoate piles of banished stone, foundations were the initial, tangible expression of the plan of a building, the base from which structure, mass, and space were generated. To ignore foundations is to ignore a crucial chapter in the history of a building and, more broadly, in the history of architectural technology. We will not fully have penetrated the mystery of Greek temples, Roman public buildings, or Gothic cathedrals until we understand the various ways in which builders laid the foundations for their support.

Bibliography

Adam, Jean-Pierre. *La construction romaine, Matériaux et techniques.* Paris, 1984.

Alberti, Leon Battista. *On the Art of Building,* trans. J. Rykwert, N. Leach, and R. Tavenor. Cambridge, MA, 1988.

Audouy, Michel, et al. *Excavations at Brixworth 1981–1982,* and *The Petrological Survey of All Saints' Church.* Brixworth Archaeological Research Committee: Two Interim Reports, 1985.

Bonde, Sheila, Clark Maines, and Robert Mark. "Archaeology and Engineering: The Foundations of Amiens Cathedral." *Kunstchronik,* 42 (1987), pp. 341–348.

Bonde, Sheila, and Clark Maines. "Entre les textes et la terre: la fouille de la salle capitulaire à Saint-Jean-des-Vignes, Soissons." *Archéologie Médiévale,* forthcoming.

Bussby, Canon Frederick. *Winchester Cathedral 1079–1979.* London, 1979, pp. 256–259.

Bussby, Canon Frederick. *William Walker: The Diver who Saved Winchester Cathedral.* Winchester, 1987.

Camp, John McK., II, and William B. Dinsmoor. *Ancient Athenian Building Methods. Excavations of the Athenian Agora.* Picture Book No. 21, American School of Classical Studies at Athens. Meriden, 1984.

Conant, K. J. "Medieval Academy Excavations at Cluny, The Season of 1928." *Speculum,* 4 (1929), pp. 3–26.

Conant, K. J. "Medieval Academy Excavations at Cluny, The Season of 1929." *Speculum,* 6 (1931), pp. 3–14.

Coulton, J. J. *Ancient Greek Architects at Work: Problems of Structure and Design.* Ithaca, 1977.

Deneux, Henri. *Dix Ans de Fouilles dans la cathédrale de Reims: 1919–1930.* Reims, 1944.

Doppelfeld, Otto. *Der unterirdische Dom.* Cologne, 1948.

Durand, Georges. *Monographie de l'église cathédrale Notre-Dame d'Amiens.* 2 vols. Amiens, 1901–1903, I, pp. 202–204.

Emerson, W., and R. L. Van Nice. "Hagia Sophia, Istanbul: Preliminary Report of a Recent Examination of the Structure." *American Journal of Archaeology,* 47 (1943), pp. 403–436.

Harvey, J. H. *The Medieval Architect.* New York, 1972.

Johnson, S. *Late Roman Fortifications.* Totowa, NJ, 1983.

Kerisel, Jean. *Down to Earth: Foundations Past and Present: The Invisible Art of the Builder.* Rotterdam/Boston, 1987.

Kurmann, Peter. *Meaux Cathedral.* Paris, 1971.

Lamprecht, Hans-Otto. *Opus Caementitium: Bautechnik der Römer*. Cologne, 1985.

Mainstone, Rowland. *Developments in Structural Form*. Cambridge, MA, 1983, esp. pp. 174–177.

Mango, Cyril. *Byzantine Architecture*. New York, 1976.

Martin, Roland. *Manuel de L'Architecture Grecque*. Paris, 1965.

McGee, J. David. "The Early Vaults of Saint-Etienne, Beauvais." *Journal of the Soc. of Architectural Historians,* 45 (1986), pp. 20–31.

Mitchell, J. K., V. Vivatrat, and T. W. Lambe. "Foundation Performance of the Tower of Pisa." *Journal of the Geotechnical Division, Proceedings ASCE,* 103, GT3 (March 1977), pp. 227–249, and "Discussion" by G. A. Leonards, 105, GT1 (January 1979), pp. 95–105.

Ousterhout, Robert. "The Byzantine Church at Enez: Problems in Twelfth-Century Architecture." *ABSA,* 1983, pp. 261–280.

Ousterhout, Robert. *The Architecture of the Kariye Camii in Istanbul*. Washington, DC, 1987.

Palladio, Andrea. *The Four Books of Architecture,* trans. Isaac Ware. London, 1738; reprint, New York, 1965.

Panofsky, Erwin. *Abbot Suger and St. Denis*. Princeton, 1979.

Paris, Bibliothèque Nationale, ms. lat. 17716.

Phillips, Derek. *Excavations at York Minster—The Cathedral of Archbishop Thomas of Bayeux,* II. London, 1984.

Plutarch. *Life of Kimon,* trans. A. Blamire. London, 1989.

Sear, Frank. *Roman Architecture*. Ithaca, 1982.

Serlio, Sebastiano. *Tutte opere d'architettura e prospetiva di Sebastiano Serlio*, 1619; reprint, Venice, 1964.

Shelby, Lon. "The Practical Geometry of Medieval Masons." *Studies in Medieval Culture,* 5 (1975), pp. 133–144.

Terzaghi, K., and R. B. Peck. *Soil Mechanics in Engineering Practice,* 2d ed. New York, 1967.

Theocharidou, K. "The Structure of Hagia Sophia in Thessalonike." In R. Mark and A. S. Cakmak, eds., *Hagia Sophia from the Age of Justinian to the Present*. New York, 1992, pp. 83–99.

Toynbee, J., and John Ward-Perkins. *The Shrine of St. Peter and the Vatican Excavations*. New York, 1957, p. 197.

Viollet-le-Duc, Eugène. *Dictionnaire Raisonné de L'Architecture Française du XIe au VIXe Siècle,* 10 vols. Paris, 1854–1868, esp. "Construction," vol. IV, pp. 175–177.

Vitruvius. *The Ten Books of Architecture,* trans. M.H. Morgan. New York, 1960.

Wulzinger, Karl. *Byzantine Baudenkäler zu Konstantinople*. Hanover, 1925.

3

Walls and Other Vertical Elements

As the zone that normally carries the decorations of church or civic architecture—frescoes, tapestries, carved capitals, and friezes, or stained glass panels—the wall prominently announces the narrative and symbolic meaning of a building. Yet any decorative or symbolic program must exist within the functional requirements of the wall, including support, access, and lighting. This chapter, dealing mainly with walls, will also treat more complex systems of vertical structural elements: piers, arcades, and buttresses, as well as systems contained within walls, including galleries and passages. All of these elements constitute the supporting connection between the foundations below and the vaults, domes, and roofs above.

Walls serve two main functional roles: to form an envelope providing security and shelter from sight, wind, rain, and temperature, and to support the weight of the building superstructure. **Load-bearing walls** combine both of these functions, acting as a continuous support to carry roof loads all along their top and to transfer them down directly to foundations. Such walls tend to be equally strong along every point of their length and are therefore usually characterized by planar surfaces and substantial thicknesses. Openings for windows and doors generally remain modest so as not to disrupt the structural continuity of the system. Walls constructed of stone, brick, and adobe normally fall into the classification of continuous, load-bearing walls.

In **non-load-bearing wall** systems, roof and floor loads are supported on vertical shafts and, typically, a lighter material fills the openings between

3.1 *Hagia Sophia, Istanbul, 532–537: interior, north wall.*

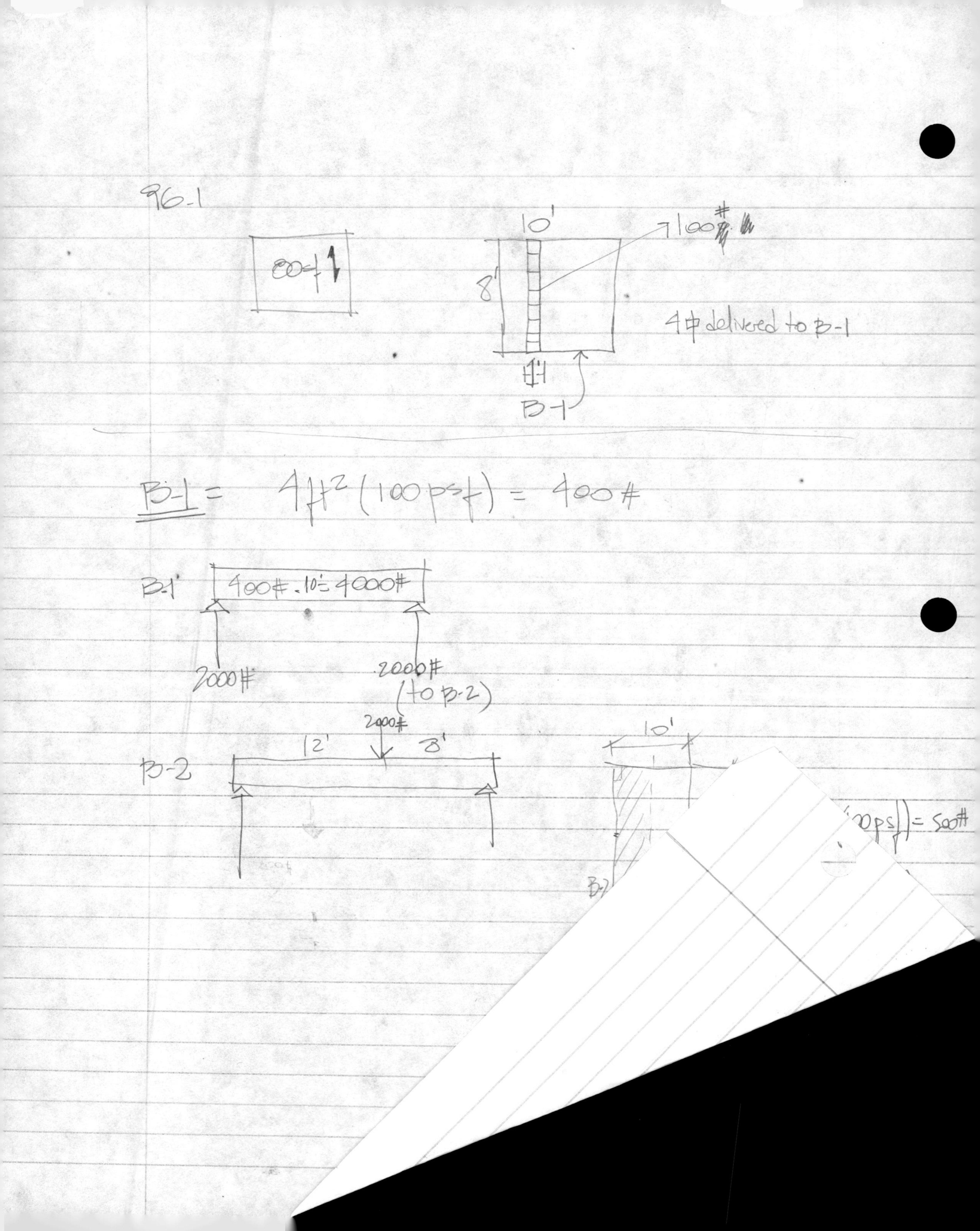

96-1
10'
8'
100#
4# delivered to B-1
B-1
B-1 = 4 ft² (100 psf) = 400#
B-1
400# · 10' = 4000#
2000#
2000#
(to B-2)
2000#
12'
8'
B-2
10'
= 500#

علي
رضي الله عنه

3.2 *Imperial Roman Basilica, Trier, early fourth century: south flank.*

the shafts. Two materials are usually employed for these walls because of the very different physical requirements for structural support and for environmental control. In half-timbered wall construction, for example, the load is transferred from the roof to the foundation through heavy timber posts, while the wall between the timbers is composed of brick or stone nogging, stuccoed over to provide a weather-tight surface. A non-load-bearing wall can be relatively thin and pierced with large windows as are, for example, the great lateral walls, or tympani (figure 3.1), beneath the massive arches supporting the dome of the Hagia Sophia in Constantinople (modern Istanbul). More common examples of non-load-bearing walls are found in Gothic churches, where loads are directed from both the roof and the vaults to points on the piers and exterior buttresses, but rarely to the walls that are then opened up to great windows and wide arcades.

Where windows occur at regular intervals in an otherwise load-bearing wall, an intermediate structural system results. In these, the vertical wall between the windows acts similarly to an isolated structural shaft; the larger the windows, the more the system approaches a fully non-load-bearing wall. This wall type is exemplified in the imperial Roman *basilica at Trier*, where the recession of the masonry spandrels behind the wall plane clearly expresses their nonsupportive role (figure 3.2).

Another hybrid system closer in spirit to the non-load-bearing wall results where the wall becomes thicker at intervals to accept concentrated loads from vaults or roofs. These projecting elements, known as wall buttresses (figure 3.3), generally coincide with the bay system of interior vaulting or the spacing of the principal trusses of the roof

3.3 *Abbey Church, Jumièges ca. 1067: wall buttresses on the north wall of the nave.*

above. The surface of wall in between the "strong" shafts needs only to support its own deadweight as well as relatively low loadings from the roof. And as the wall thickness between the buttresses decreases, this system, too, approaches the skeletal support so evident in mature Gothic design.

Building loadings may be classified as *dead* or *live,* depending on whether they change with time. Dead loads derive from the fixed mass of a building's structure, while live loads are caused by time-dependent external factors such as wind, earthquake, and the motion of people and furniture within the building (figure 3.4). In a simple wall, the action of the dead loads alone usually results in a state of pure compression, as illustrated in figure 3.5. When any material is compressed, including stone, it compacts in a similar manner to a squeezed sponge. Stone, of course, is much stiffer than sponge, and such changes cannot be observed by the naked eye. In fact, no building material is absolutely rigid, but some materials behave relatively rigidly compared to others. A structural element composed of iron or of steel, for example, is some ten times stiffer (i.e., it will deflect only one tenth as much) than the same element made of stone, and about twenty to thirty times stiffer than the same element made of construction-grade timber.

Under ordinary, short-time loading conditions, most building materials can be considered *elastic*, that is, when the loading is removed, they return to their original form, as does a rubber band. Extreme loadings, and loadings of long duration, produce additional permanent deformation, called *creep.* In modern engineering practice, when two materials with widely differing stiffness, such as stone and timber, are used together in construction,

3.4 Structural loadings.
To determine the deadweight gravity loadings *acting within a structure, one needs first to calculate the volumes of material and the locations of the centers of gravity of the individual building elements—usually from detailed drawings of the building, but often supplemented by on-site measurements. The magnitudes of the loadings are then found by multiplying the volumes by a standard unit weight for the particular material; for example, the unit weight of construction stone is generally taken as 2,300 kg/m*^{3}.

For estimating the wind loading *on a tall building, one must first consult local meteorological records for the general wind speeds and directions over extended periods of time, as well as theoretical wind-velocity profiles (velocity vs. height above ground level) for the particular terrain of the building site. Wind-velocity data over long periods of time, even as long as a century, can usually be obtained from governmental meteorological sources. Maximum winds normally occur over a fairly wide azimuth, so the full wind loadings are considered to act in their most critical direction, usually transverse to a building's longitudinal axis. Wind-pressure distributions (and suction on the leeward side of a building) are then calculated from these data and from wind-tunnel test data for the particular configuration of the building by means of the equation*

$$p = \tfrac{1}{2}\beta \times V^2 \times C \times G,$$

where p is the wind pressure at any point on the building surface, β *is the mass density of air; V is the wind speed; C is a dimensionless coefficient related to building form, usually established from wind-tunnel tests; and G is a gust factor to account for the dynamic action of impinging air (Mark 1982, 22–25). A typical middle European preindustrial townscape wind speed and pressure distribution (with C = G = 1.0) is illustrated. Note the high sensitivity of pressure to wind speed; doubling the speed gives four times the wind pressure.*

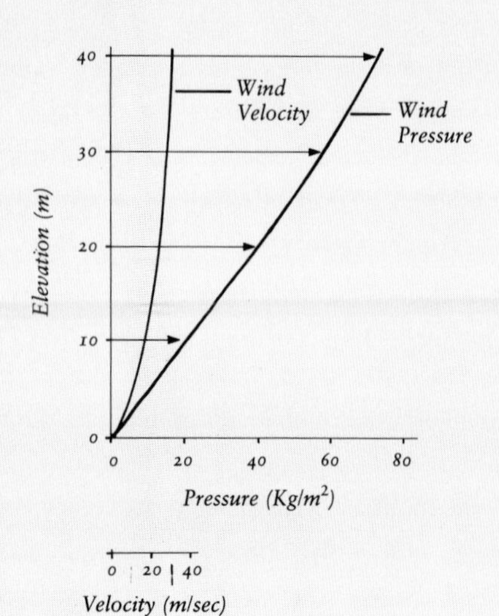

Earthquakes are essentially vibrations of the earth's crust accompanying dynamic adjustment to subterranean ground faults. During an earthquake, seismic loadings *are induced as the ground surface moves in all directions, and inertia causes the building to resist these motions. Usually the most perilous ground motion for buildings is horizontal, along the ground surface, which generates lateral forces, similar to those caused by wind, throughout the structure.*

3.5 Wall forces from deadweight.
For a loaded structure to maintain its integrity (equilibrium), resisting forces within the structure must counteract the applied loadings. Pulling on a sapling, for example, subjects the sapling to tension *(a stretching force) of the same magnitude as the applied force. Similarly, the illustrated wall undergoes* compression *(a pushing force) from its own weight. At the top of the wall the compression force is zero; at the base, the compressive force equals the total weight of the wall.*

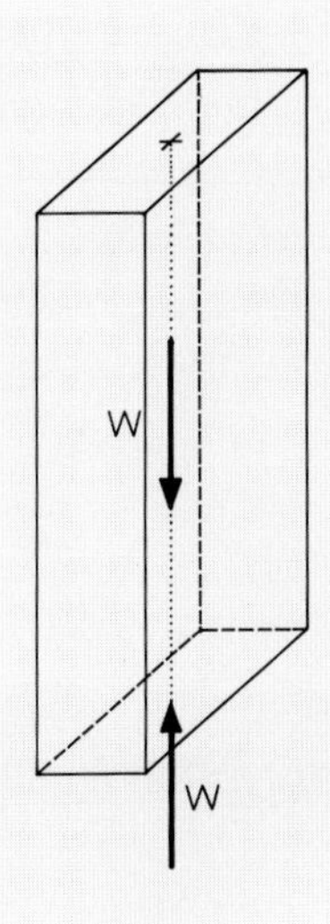

3.6 Wall forces from combined deadweight and applied loading; reactions; overturning.

The inclined, applied force F *acting on the top of the wall (dashed line) which, for example, represents thrust applied by a groined vault, can be* resolved *into vertical and horizonal* components; *that is, the inclined force can be replaced by two imaginary forces that together have the same effect as the single force. These are found geometrically: If the length of line* F *represents the magnitude and direction of the inclined force, the lengths of the vertical and horizontal legs of the right triangle formed with* F *as hypotenuse give the magnitudes of these components. The vertical force component* V *adds to the* compression *from the wall's weight, and the horizontal force component* H *subjects the wall to internal* bending *and* shear *forces. Under bending, the side of the wall meeting the load stretches; and the opposite side experiences additional compression from bending. A combination of all three internal forces is usually present in any structure.*

Reactions *are the internal forces that provide support to a structure. For example, the supporting reaction at the base of the wall illustrated in figure 1.5 is a vertical compressive force equal to its total weight. With the addition of the inclined force,* F, *there must be three reactions: a vertical compression equal to the sum of the weight of the wall and the vertical force component* V; *a bending force (or "moment," similar to torque exerted by a wrench) equal to the product of the horizontal force component* H *and the height of the force above the base,* y; *and a shearing force of magnitude* H.

Overturning *occurs after the base section is cracked and the applied bending force* (H × y) *exceeds the "righting moment"* (W + V) × (t/2) *set into play by the downward forces tending to rotate the wall oppositely about its outside edge. Hence raising the wall (to increase its weight,* W) *or splaying out the wall base (to increase its thickness,* t) *helps to stabilize it.*

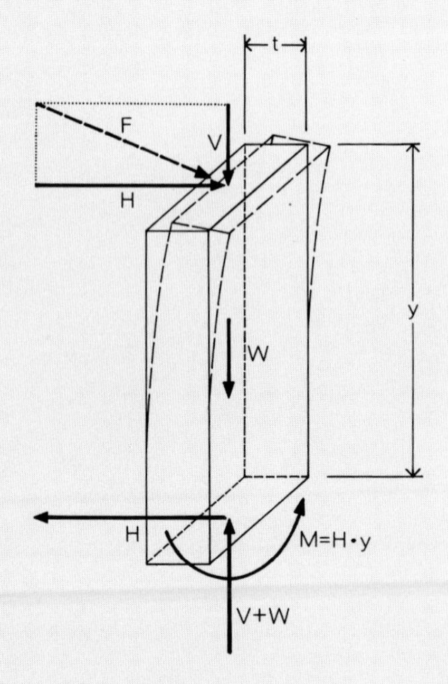

one usually assumes that the stone remains rigid and that the deformation takes place mainly in the timber.

Before the Scientific Revolution, almost all western monumental buildings used stone for their structural walls and, even in even the tallest of these buildings, the high compressive stress levels almost never resulted in failure. Part of this history of success is due to the layer of mortar between the stones that helps to distribute compressive forces over the full contact surface, rather than allowing the structure above to rest on a few high points in the cut stone. Moreover, stone is very strong in compression. Medium sandstone, for example, which is not particularly strong and weighs about 2,300 kg/m^3, boasts a crushing strength of about 425 kg/cm^2. The Washington Monument, the tallest unreinforced masonry tower in the world, is essentially a hollow shaft with walls increasing in thickness toward the base. Yet even if the shaft had no taper and was instead built of solid masonry, its 171-meter height would generate but 40 kg/cm^2 of compressive stress at its base due to vertical deadweight—less than 10 percent of sandstone's crushing strength.

On the other hand, tension, occurring when materials are pulled apart or stretched, is considerably more perilous for masonry. While some construction materials, such as wood and iron, have nearly equal strength in resisting tension and compression, stone and brick are relatively weak in tension compared to their strength in compression. Even slight levels of tension in a masonry wall can result in cracking. A single block of stone will usually display appreciable tensile strength, but mortared joints cannot be depended on, over time, to transmit tensile forces reliably.

Walls experience **bending,** or in extreme cases, overturning, when they are subject to lateral loadings. The main sources of such loadings are wind, earthquake, and the lateral thrusts of internal arches or vaults (see figure 3.4). For a wall subject to only light vertical loads but relatively heavy lateral loads, as might be experienced by an infilled wall supported by timber posts in a high wind, overturning is usually averted by some sort of lateral bracing—furnished in many buildings by internal cross-walls. In some large masonry buildings without cross-walls, however, stability against overturning from lateral loads depends on the massive deadweight of the wall itself (see figure 3.6) or upon external braces, such as the flying buttresses illustrated in figure 3.7.

Structural elements subject to bending experience far more complex distributions of stresses than those from pure compression, as illustrated in figure 3.8. Bending of a pier or wall causes one face to shorten and the other to elongate. The stretched material then experiences tensile stress that can lead to cracking, especially in the layers of mortar. As already noted, it would be a mistake to consider masonry walls to be securely "cemented together." Mortars serve mainly to avoid stress concentrations at stone and brick interfaces and provide sealing for the joints.

Fortunately, a state of pure bending is rarely encountered in masonry construction. The bending caused by lateral loadings is usually accompanied by compression from the deadweight of masonry in the structure above; and in almost all in-

3.7 (Overleaf) *Bourges Cathedral, choir, 1195–1214: flying buttresses.*

3.8 Wall stresses.

(a) From axial forces alone (axial stress): Stress is a measure of the local intensity of force acting within a structure. For a simple, concentrically loaded structure such as the sapling described in the caption of figure 3.5, the axial, tensile stress *is found by dividing the total force applied to the sapling by its cross-sectional area. For example, a 10 kg force applied to a sapling having a cross-section area of 0.1 cm^2 produces a tensile stress of 10 kg/0.1 cm^2 = 100 kg/cm^2. In the same manner, axial compression forces gives rise to uniform* compression stress. *The compressive stress at the base of the wall of figure (a) is similarly found by dividing the total co-axial vertical load* (W + V) *by the cross-section area (A) of the wall base. (b) From bending alone* bending stress*: The magnitude of the* bending stress *accompanying moment (caused by the lateral loading,* H) *is inversely proportional to the product of the wall thickness* t *and its cross-section area (see Gordon, 377–379). Since any increase in wall thickness results also in an increase in the wall area, the bending stress varies inversely with the square of the wall thickness ($1/t^2$). In other words, doubling the wall thickness reduces bending stress to only 1/4 of its initial value. Maximum tensile stress occurs at the surface on the side of the wall meeting the load, and it is here that cracking tends to develop in masonry. Bending stress is effectively zero at the wall center, while maximum compressive stress occurs on the opposite wall surface, as illustrated.*

A horizontal **beam** *in bending behaves in exactly the same manner as the illustrated vertical wall. Like the wall, bending stress is inversely proportional to the product of the beam depth and cross-section area. Hence, doubling the beam depth reduces bending stress to only 1/4 of its initial value.*

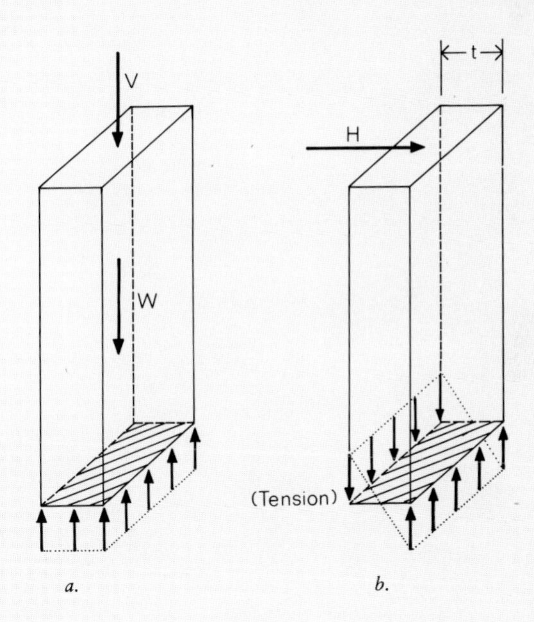

3.9 Wall stresses from combined axial and lateral forces (combined compression and bending stresses).

(a) Low bending compared to vertical loads.

(b) Moderate bending compared to vertical loads.

(c) High bending compared to vertical loads.

(d) Extreme bending compared to vertical loads; wall overturns.

Note: In addition to the illustrated, axial stresses arising from axial and lateral forces, the internal shearing force (H) *gives rise to* shear stresses *that can also cause a wall to fail through stones sliding over one another. This problem, though rarely encountered in historic buildings, can be ameliorated by raising the wall height, thereby increasing the sliding friction between the stones.*

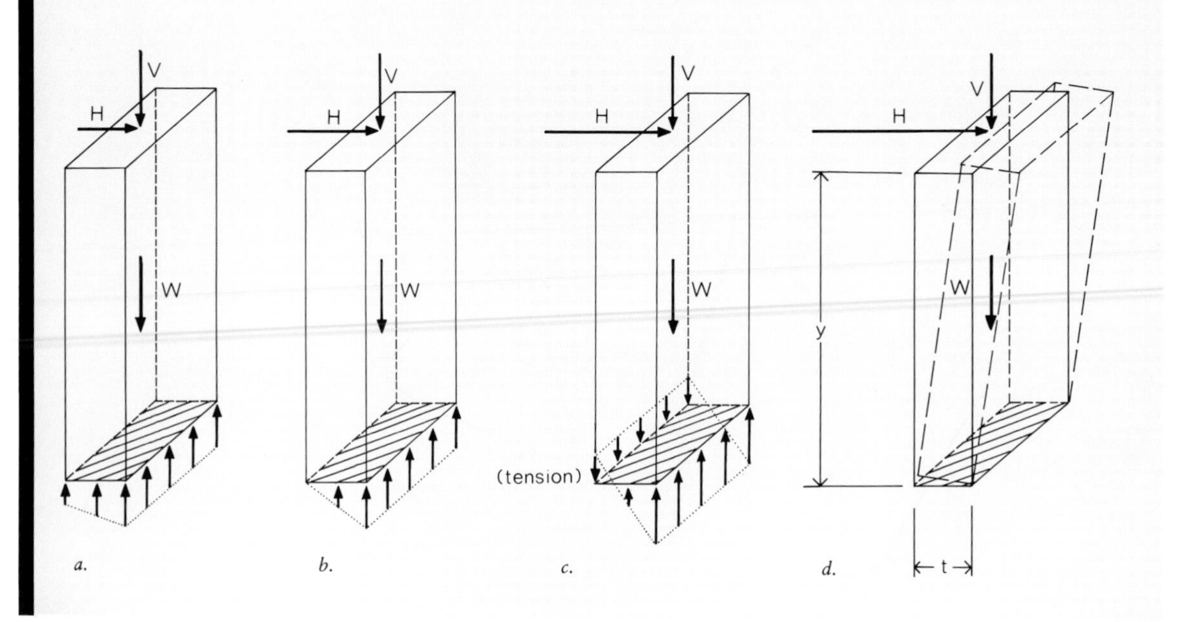

3.10 Wall stresses from offset axial force; thrust lines.

Force V, *applied at eccentricity (e) from the center-line of the wall is equilibrated by a concentric force* (V) *and a bending moment (equal to* V × e*). Hence the stress distribution corresponding to an eccentric force is the* same *as those given in figures 3.8a and 3.9 for axial, and combined axial and bending forces.*

In the absence of lateral forces, the so-called, thrust line, *or resultant of axial forces such as* W *and* V *(in figure 3.8a), runs vertically down the wall center (i.e.,* e = 0*) and the resulting stress distribution is given by figure 3.8a. When lateral loads such as those due to vaulting or wind push against the wall, the thrust line is displaced off center, with its eccentricity (e) increasing in proportion to the magnitude of the lateral loads. In extreme cases, where the effective thrust line falls outside the wall or pier face (i.e.,* e *is larger than* t/2*), failure results as in figure 3.9d because of the inability of most mortars to reliably "cement" the stones together. A more common, and less dire, problem is experienced when the thrust line falls within the face of the wall, but outside of the so-called* middle third *(i.e.,* e *is greater than* t/6 *but less than* t/2*). For this range of loadings, highest compressive stresses occur on the face of the wall, or pier, closest to the thrust line, and tension, which generally results in masonry cracking, is displayed on the far side. With the thrust line within the boundary of the middle third of the wall or pier (i.e.,* e *equals or is less than* t/6*), stresses throughout the section are compressive (figures 3.9a and b) and cracking will not develop (see Schodek, 288–290).*

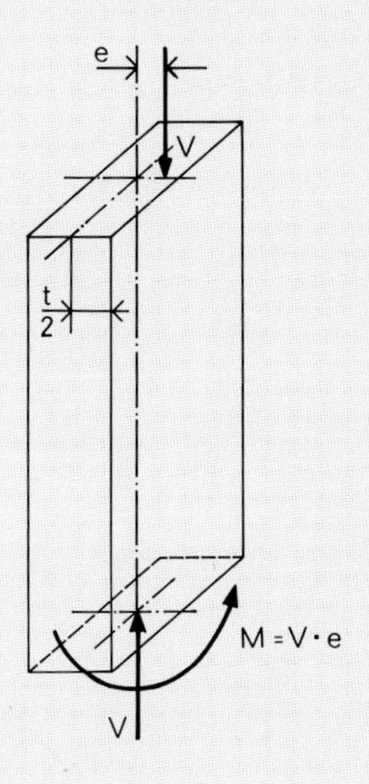

stances, both surfaces of a wall will experience compression, as illustrated in figures 3.9a and 3.9b. When tension does develop, as illustrated in figure 3.9c, the ensuing cracking may not in itself prove destructive to the integrity of a structure. Yet cracks can allow the entry of water, which can wash away lime mortar, or in the presence of freezing temperatures, expand and further break up the masonry. Although both phenomena can precipitate great reductions in wall strength, master masons could make design corrections by examining their structures for the presence of such cracks, even during the process of construction. The thickness of a wall or pier can be increased, or they might be made taller to increase compression forces that tend to close up the cracks, an especially attractive option when foundations have already been completed and the width of the wall fixed. Heavy parapets or cornices were sometimes been used to solidify walls, as have pinnacles and even statues (Mark 1990, 118–120). Even so, it is highly unlikely that master masons understood the theory of the rule of the "middle third" defined in figure 3.10.

Although many problems of structure could be avoided simply by constructing thick, heavy walls or piers, costs escalate with the greater volume of stone to be quarried and transported, not only for the wall itself but also for the foundation providing its support. When lateral loads are applied at localized regions of the wall, such as at the supports of groined vaults or below domes supported by arches and pendentives, it is more economical to thicken the walls locally to form a wall buttress. Another viable structural solution involves inclining the wall or pier to match the angle of the resulting force, as with a steeply pitched flying buttress connecting the springing of a vault to a free-standing pier. The flying buttresses of the cathedral of Bourges (figure 3.7) stunningly illustrate how slender and lightweight such a structure can be.

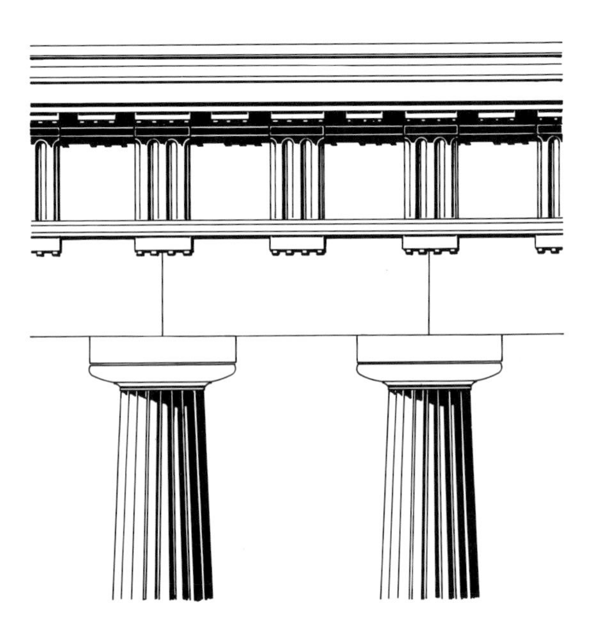

3.11 *Lintel support from column capitals.*

The simplest method for creating the openings in walls necessary for doors and windows is through the use of trabeated supports, that is, **post and lintel construction.** In this system the jambs, or sides of the openings, act as supporting posts, and the lintel over the top acts as a beam in bending. Wood makes for an efficient lintel because of its good tensile strength and light weight; but because of the greater relative durability of stone, wooden lintels were rarely used in monumental buildings. On the other hand, stone, being weak in tension, makes for a most inefficient lintel. Openings framed with monolithic stones are therefore limited in span.

Ancient architects were well aware of this limitation and developed schemes to circumvent it. One method of compensating for stone's poor per-

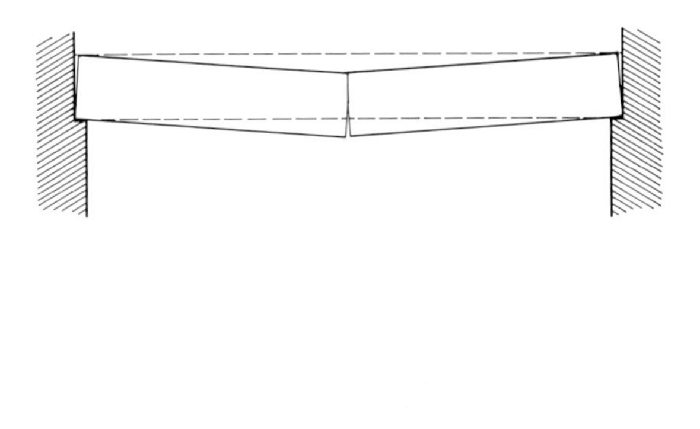

3.12 *A deep lintel, cracked at three points, remains stable if its supports are unyielding.*

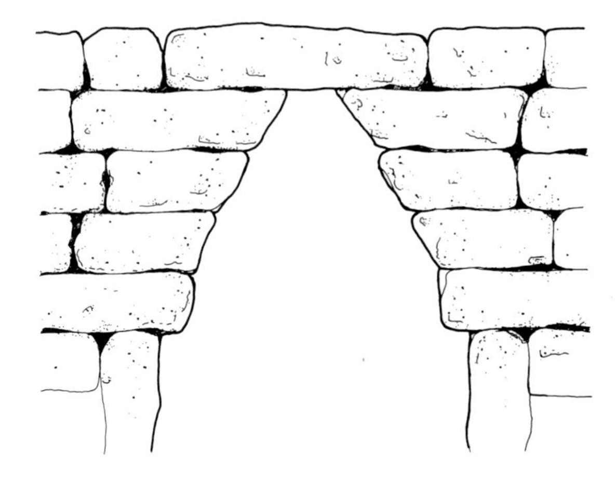

3.13 *Corbel-arch figuration.*

formance in bending is to reduce the effective clear span of an opening by employing wide capitals on top of columns (figure 3.11). Such a small reduction in clear span may seem inconsequential, but since bending stresses in a lintel vary with the square of its unsupported length, even small increments in span can have a major effect on lintel stresses. Furthermore, cracking of a stone lintel does not necessarily lead to its destruction. If the ends of a cracked, deep lintel are prevented from spreading apart, it will not be able to collapse, as illustrated in figure 3.12. A cracked structure is, of course, more susceptible to seismic movements; and freezing water in colder climates can enter the cracks and force the segments of the lintel apart. Yet many ancient, cracked lintels have survived for centuries in southern Europe and northern Africa (Heyman).

Though not so practical for providing openings in walls as the trabeated system, the so-called **corbel arch** averts many of its problems. While large spans in a trabeated building necessitate the use of correspondingly large and unwieldy monolithic lintels, corbeling can be constructed from relatively small elements, usually of cut stone or brick, each of which projects into the opening slightly past the element beneath it, as illustrated in figure 3.13. And unlike true arches, corbels require no supporting centering in the construction process; the stability of the individual elements is assured by the mass of wall placed above. But because it does not act as a true arch, an important limitation of a corbeled opening is that its height must be far greater than its base width. As with the lintel, this limitation effectively restricts practical span lengths.

True arches, however, circumvent the intrinsic span limitation of the trabeated system and the corbel arch by securing all of the constituent elements, known as *voussoirs*, in a state of compression. In effect, the curvature of an arch, as opposed to the linear form of a lintel, engenders horizontal as well as vertical reactions (figure 3.14a). The forces generated within the arch by these reactions then act

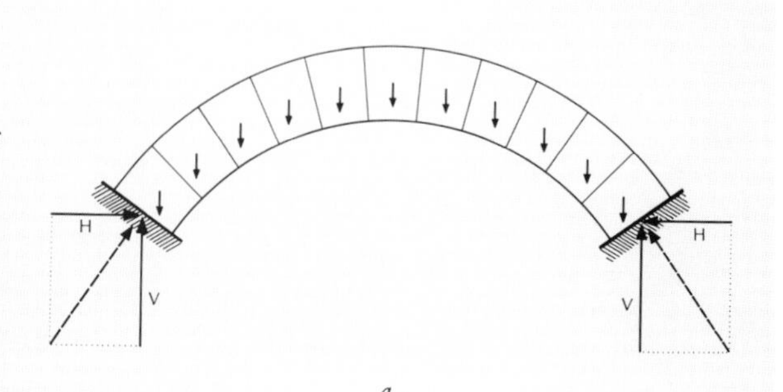

a.

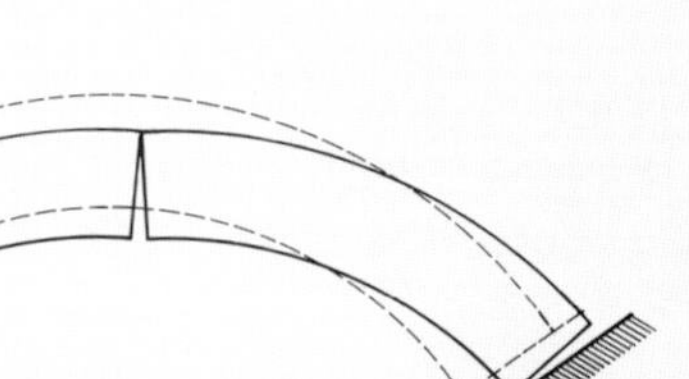

b.

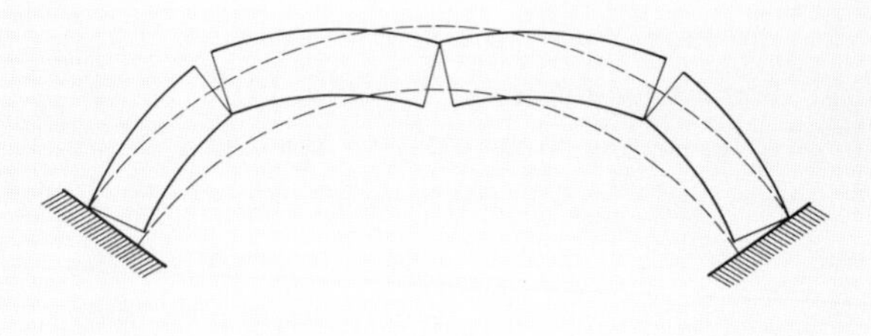

c.

3.14 True arch behavior.
(a) Form, loading, and reactions.
(b) Abutment failure. With the first spreading of an abutment, a masonry arch will likely acquire three "hinges" due to cracks forming at both abutments as well as at the crown. Nonetheless, a three-hinged arch is stable and will likely endure unless the motion of the abutment becomes greater. This characteristic of arches often allows for the reinforcement of an insubstantial abutment before major damage ensues.
(c) Four-hinge failure mode (after Coulomb).

3.15 *Segovia Aqueduct, first century A.D. Detail of the upper arcade with characteristic surcharge.*

to confine the voussoirs. Masonry arches afford great interior spans (reaching 33 meters by the sixth century; see comments on Hagia Sophia, below), a potential that established the arch as the structural system of preference for large-scale monumental buildings. True arches also share a major constructional advantage of the corbel system: assembly from relatively small, easily managed elements.

Yet despite all of these intrinsic advantages, arches demand special treatment in construction. Because they produce horizontal reactions from vertical, gravitational loadings, arches require either rigid abutments or tension ties across their bases to prevent spreading and possible collapse, as described in figure 3.14b. Moreover, an arch may prove unstable (and fail in the mode illustrated in figure 3.14c) unless its form approaches a singular, optimal shape (a catenary for a uniform arch supporting its own deadweight, but closely enough approximated by a parabola or even a shallow circular arc for most practical arch applications) and/or it is thick enough to maintain the resulting thrust line within the confines of its extrados and intrados. (The concept of thrust lines is discussed in figure 3.10). More typically in early structures, the haunches are surcharged (provided with heavy fill) to prevent large displacements of these regions of the arch (see figure 3.15).

When an arch is inserted into a wall, the region of the wall above the arch acts as surcharge. Moreover, when an arcade of arches is placed in a wall, the horizontal thrust from each arch counters that of its neighbor, so that the supporting pier below experiences only vertical compression. It only remains to securely anchor the ends of the arcade. For this reason, in Roman aqueducts, massive piers were positioned at any point where the aqueduct changed direction (figure 3.16). In large medieval churches, the relatively heavy crossing piers and the semicircular (in plan) apse usually buttress the ends of choir arcades, while the arcades of the nave are buttressed by the remaining crossing piers and the twin towers of the facade.

One of the more important questions in arch (as well as vault and dome) construction con-

3.16 *Segovia Aqueduct: buttressing at junction.*

cerns the temporary **centering** used to support voussoirs until the placement of the final voussoir, or keystone. Since it had to be designed to be taken down after construction, often for reuse, we have little first-hand knowledge of the centering that was used to support both workman and stonework during the erection process. Although in its most primitive form, centering involved the use of tamped, mounded earth (Fitchen, 30–31), timber was the primary medium employed for formwork in historic buildings and therefore, both structurally and economically, it constituted an essential element of masonry construction. Precision of execution, rigidity of form, and the ease of removal of this temporary structure played a key role in the building process.

The centering, which determines the profile of the underside of an arch (or a vault, as discussed in chapter 4) remains in place until the completed arch can stand on its own. While in use, though, the centering must resist deformation as incremental loadings are applied, to keep the desired final shape of the arch. Hence, the centering used for large-scale construction, such as in imperial Roman architecture, was of necessity powerfully built.

Centering falls into two basic types: (1) that supported directly on the ground, by means of vertical or radial wooden struts, and (2), that springing from a masonry pier, wall, or vertical support at the end of an arch (figure 3.17). The second method is especially expedient for bridge construction, as the river flow might wash out temporary columns placed in the main channel. It also saves timber whenever the arch is being constructed high above the ground, and so was employed in the majority of these cases. The technique was used, for example, in the construction of the Roman aqueduct at Nîmes, the Pont du Gard (figure 3.18), where one can still observe three levels of masonry projections from which the centering was sprung (Adam, 191). Note that building up the haunches of each side of the arch and "flying" the centering from both the column capital and uppermost part of the springing, as in figure 3.19, reduces the actual *arc* to be built to a segment smaller than a half-circle, thereby saving labor and materials, thanks to the reduction in centering span. In addition, this order of construction ensures that the surcharge has already been placed prior to decentering, so that the haunches of the arch have no opportunity to rise and deform.

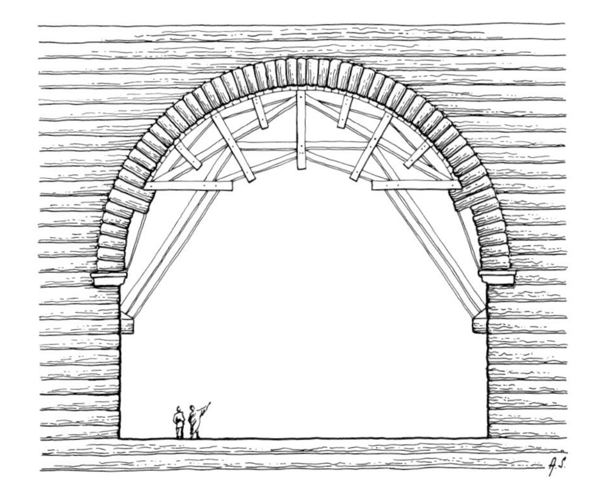

3.17 *"Flying centering" used in Roman bridge construction.*

3.18 *Pont du Gard aqueduct, Nimes, first century,* A.D.

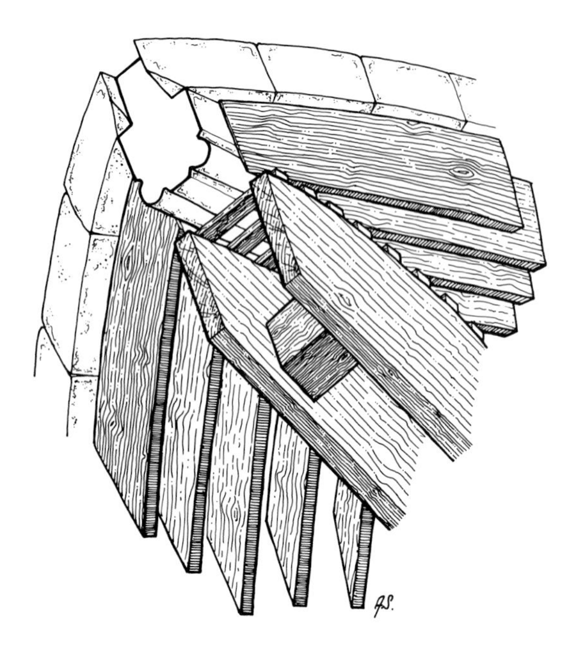

3.19 *Timber centering supporting ribbed-vault construction (after Fitchen).*

In its simplest form, when used to erect a single, semicircular masonry arch, the centering timberwork consists of at least two parallel arches braced by triangulated framing. These arches, made up of short, joined timbers, carry between them planks, known as laggings (illustrated in figure 3.19), upon which the stones are set. To provide for safe decentering, builders inserted pairs of opposing wedges beneath the wooden centering either on temporary foundations or, in the case of flying centering, on masonry projections. Upon completion of the arch the centering was struck by driving out the wedges, which in turn dropped the centering a few centimeters and allowed the voussoirs to wedge themselves into place. In large arches with many wedged supports, decentering was a difficult procedure that required close coordination in the removal of wedges, as uneven decentering might result in distortion or collapse of the structure.

Many approaches were employed to reduce the expense of centering, the most common of which was reuse. Wide masonry arches were often built up over a series of parallel arches. In figure 3.20, for example, three similar stone arches are clearly seen from below the intermediate arcade of the Roman Pont du Gard. With the keystone of the first arch in place, the opposing wedges were driven out and the freed centering was slid sideways and raised with wedges for the construction of the next parallel arch. As four similar arches were also used for the wider, lower arcade of the aqueduct, a single, thin centering seems to have served for the construction of all seven. Where large stone voussoirs were used, the lagging would be stoutly proportioned and widely spaced, with each voussoir bridging the gap between the lagging, as at the Roman aqueduct at Segovia (figure 3.15). Closer spacings were required in Roman concrete construction where rubble stone or brick was laid in thick beds of hydraulic pozzolan mortar. When the centering was struck, the impression of the closely spaced lagging remained, just as do formwork patterns on modern concrete surfaces.

Centering costs could also be reduced by first building a relatively light arch on correspondingly light centering, and then using the completed arch to support additional concentric rings of masonry. The Romans used brick and tile arches in this manner, building up much heavier forms on what became, in essence, permanent centering (Viollet-le-Duc, IX: 465–467). This practice was continued in Romanesque construction, especially in church portals in southwest France, where the concentric arches became an aesthetic focal point (figure 3.21).

3.20 *Pont du Gard: intermediate arcade from below.*

3.21 *Aulnay, 1119–1135: facade.*

3.22 Natural light illuminance vs. building scale.

The level of natural-light illuminance (surface brightness) at an interior point in a building is directly related to window area, the inverse square of the light-path-lengths from the windows (light levels at twice the distance from a source will be but 1/4 as strong), and the orientation of the axes of the light paths (Mark 1990, 43–47). Assuming the same source intensity of light for two similar buildings of different scale, both will experience the same *levels of illuminance because the longer light paths of the larger building serve to reduce light transmission by an inverse-square relationship while the area of its window openings, and hence its sources of lighting, are proportional to the square of its scale. The two effects thus cancel each other.*

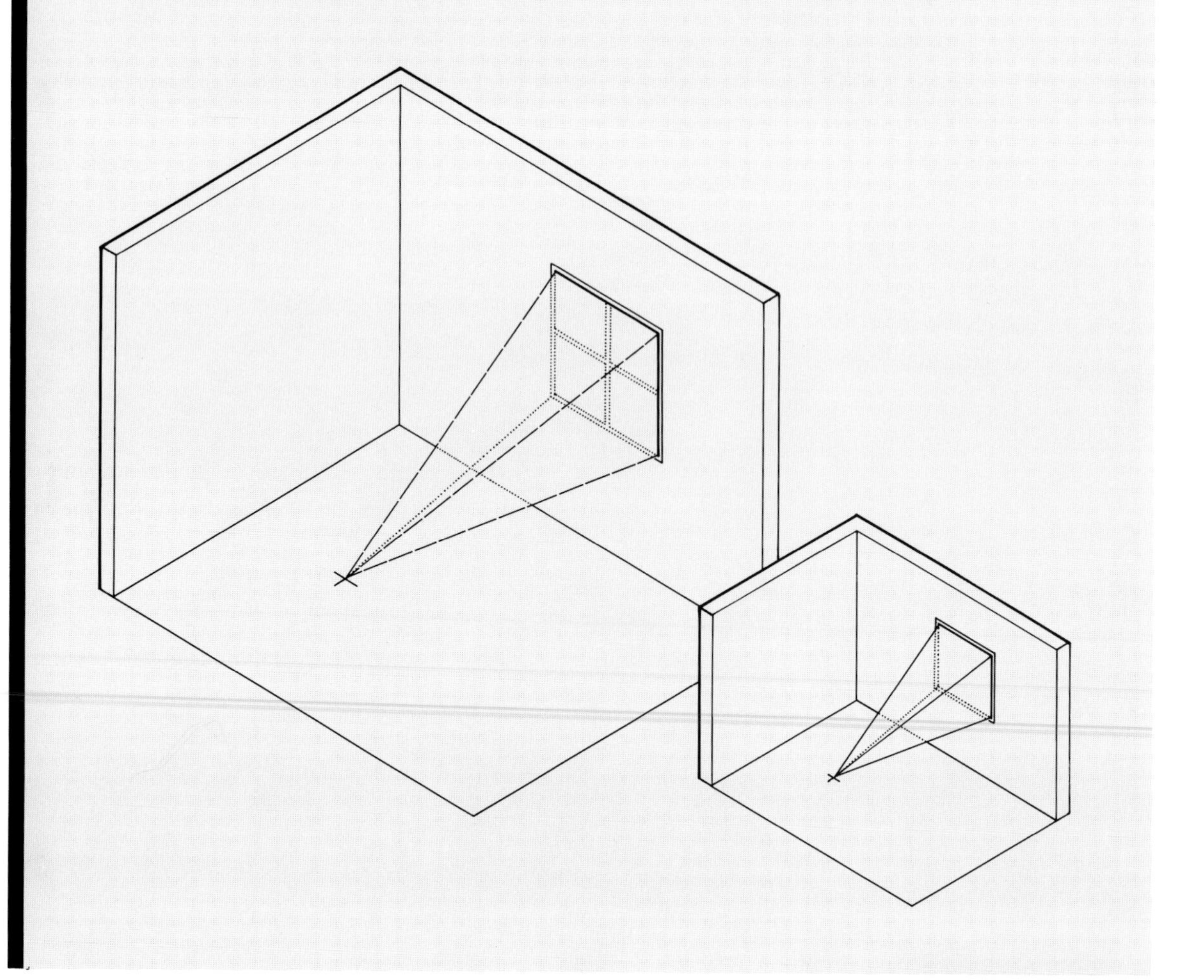

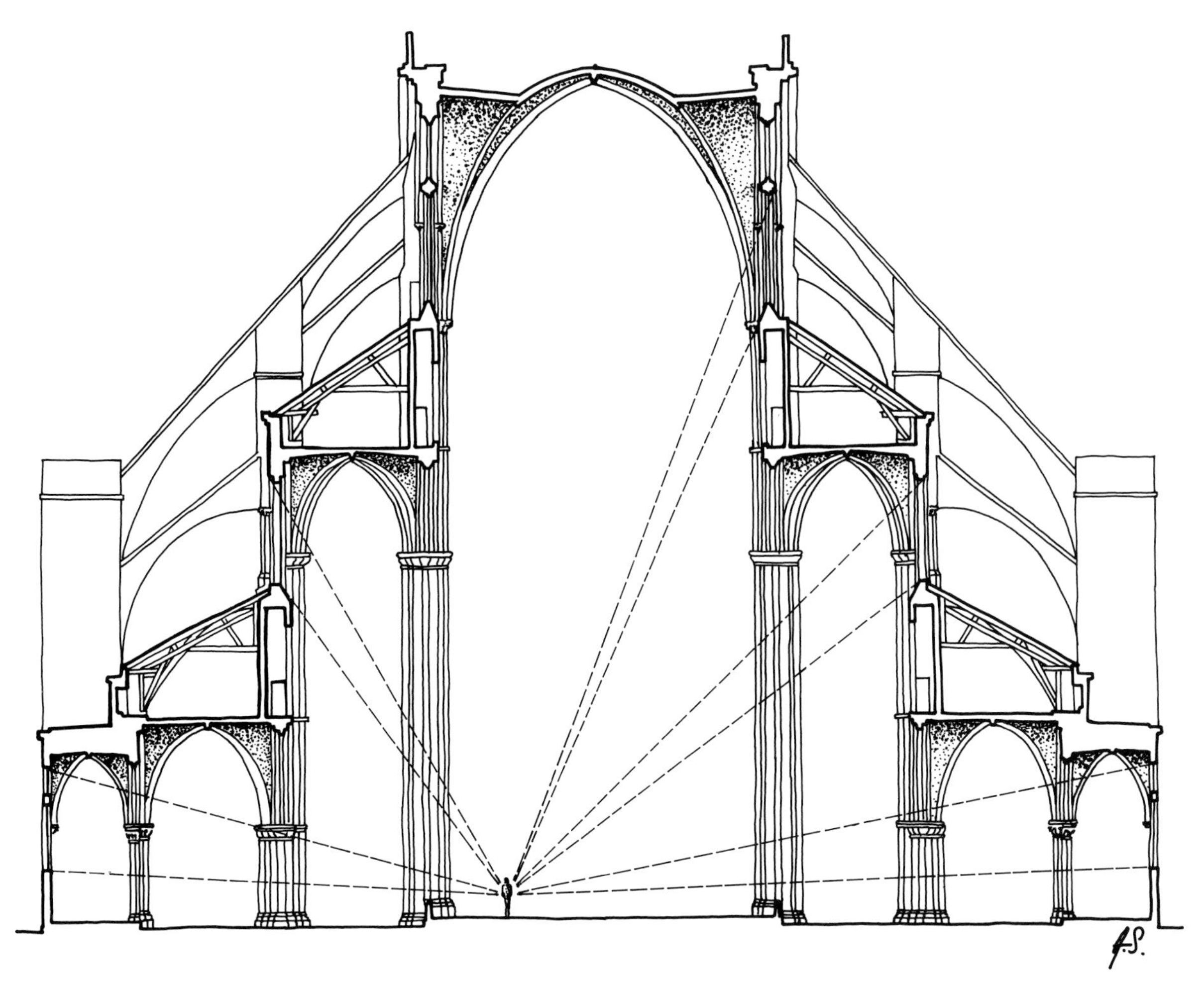

3.23 *Bourges Cathedral: cross-section of the choir showing lightpaths from wall openings to a region of the nave floor.*

Interior lighting needs also played an important role in the development of wall forms. The desire for large windows, for example, was central to the evolution of the non-load-bearing wall. Whatever the form of the wall openings, the materials adopted by different societies to enclose them have ranged from glass and fabric to mica-like minerals. In the Romanesque and Gothic periods, particularly, the wall became a surface activated both visually and structurally by a series of arched and vaulted openings, illuminated by panels of stained glass. Light admitted through these openings could be interpreted symbolically, as Abbot Suger's twelfth-century writings on the windows of St. Denis reveal. Light was considered a force for the elevation of the spirit toward God, and Suger believed that the luminous interior of his church brought the worshipper out of a purely physical realm into a higher state of contemplation (Panofsky, 23).

The creation of light-filled interiors, especially in the Middle Ages, may well have been metaphysically motivated; yet in lighting, physics plays a more direct role than metaphysics. Since historic buildings were dependent mainly upon sunlight for illumination, describing the effective source intensity at a window proves complex: it varies with the time of day, weather, season, orientation of the windows, external obstructions such as surrounding buildings and, of course, the light transmission permitted by window fittings, especially by stained glass. Although one must interpret with caution any purely quantitative data on architectural lighting, paying due attention to aesthetic and symbolic concerns, analysis of the interior lighting can in some cases help to clarify design intentions in historic buildings.

Illuminance, or surface brightness, provides the most readily accessible measure of interior lighting. As shown in figure 3.22, illuminance remains unaffected by building scale, since for larger buildings the windows are both greater in size and farther from the observer. Geometry, on the other hand, plays a major role in determining lighting levels. As the viewing angle becomes increasingly oblique, for example, the window appears more foreshortened and provides less light to the observer for a given surface brightness of the window. Consequently, windows set high in the wall lose their effectiveness in providing intense levels of illuminance to viewers at ground level, as demonstrated in figure 3.23 comparing the light reaching the floor from the upper and lower clerestories of Bourges Cathedral.

Choices of materials also affect architectural lighting conditions. Above and beyond the choice of window covering, the reflectivity of interior surfaces and the presence of artificial light sources can contribute significantly to both the amount and the quality of light. Many Greek temples, for example, depended on candles and torches to light their interior cellas. Despite all of these complications, however, satisfying a patron's taste for lighting conditions must have always constituted an important design goal for the early builder, one that also motivated notable structural innovation in wall forms.

Having examined aspects of design, construction, and the structural behavior of historic walls and other vertical elements in general terms, we may now consider the application of these principles to specific buildings constructed before the scientific revolution.

Ancient

Greek monumental buildings were almost exclusively based on trabeated wall construction. True arches were certainly known to the Hellenistic Greeks, who did in fact use barrel vaulting based on the arch, but not often, and then usually in subterranean, utilitarian structures or tombs. Earlier Greek cultures employed corbel arches, with perhaps the best known of these being the partially destroyed Lion Gate at Mycenae, constructed in the mid-thirteenth century B.C. (figure 3.24). In addition to exhibiting the corbeling technique, the gateway incorporates a heavy stone lintel deepened at its midsection (where bending is greatest—benefit from increasing beam depth is discussed in figure 3.9b) to span a 3.2 meter-wide opening.

Until the mid-eighth century B.C., Greece possessed no truly monumental architecture. Building activity seems to have increased in both scale and intensity during the following two centuries as the basic forms of the Doric temple were developed. First constructed of timber, early Doric temples were adapted to stone construction, most likely because of stone's greater durability. Even so, Greek builders had employed many different construction materials before settling on stone for monumental architecture. Low-cost walls composed of mud-brick or field stones, laid dry or in clay and reinforced with timber, appeared very early and were common even in classical times for modest buildings. The introduction of fired brick together with mortar and concrete in wall

3.24 *The Lion Gate, Mycenae, mid-thirteenth century B.C.*

construction, however, dates only to the Roman era, in the first century A.D. In most Greek monumental buildings of this later period, walls would have been covered with a thin revetment of marble.

Many elements of the **Doric order** (figure 3.25) probably represent the stylistic continuation in stone of formerly wooden elements: architraves evolved from the wooden beams spanning the columns, triglyphs and metopes arose respectively from the beam ends and from the infill between those beams, and cornices derived from the supporting member for the rafters. Yet in their newly monumental stone buildings, Greek architects confronted a new set of problems concerning materials, transport, and of course, structural support.

As early as the seventh century B.C., cut blocks of easily worked limestone were used for monumental buildings and by the sixth century Athens had begun to import fine white marble from the islands of Paros and Naxos for use in both building and sculpture. By the early fifth century, the Athenians turned to sources closer at hand and began exploiting the rich marble quarries on Mount Pentelicus (Bruno, 327). These new quarries became the primary source of building stone for Athenian religious and public buildings, and their stone was widely exported as well (Dinsmoor, 188). When lesser-quality stone was used, wall surfaces were often covered by stucco.

As the volume of stone use increased, builders devised techniques for lessening the cost of transportation. Frequently blocks were hollowed out before leaving the quarry, especially during the Archaic period when building stone often was shipped by sea from distant sources (Coulton, 146). An example of this strategy, probably because of its remote location, is found at the Temple of Apollo at Bassai (ca. 430–400 B.C.). The Parian marble ceiling beams of the temple are formed with U-shaped sections as illustrated in figure 3.26. The weight of the beams before voiding, about 2.4 tons, should have presented no unusual difficulties for lifting with a contemporary hand-powered crane (Landels, 84–85), but the approximately 40 percent reduction in weight would have been advantageous for transportation as well as in reducing the deadweight loading of the finished beam itself. The U-shape is also structurally logical

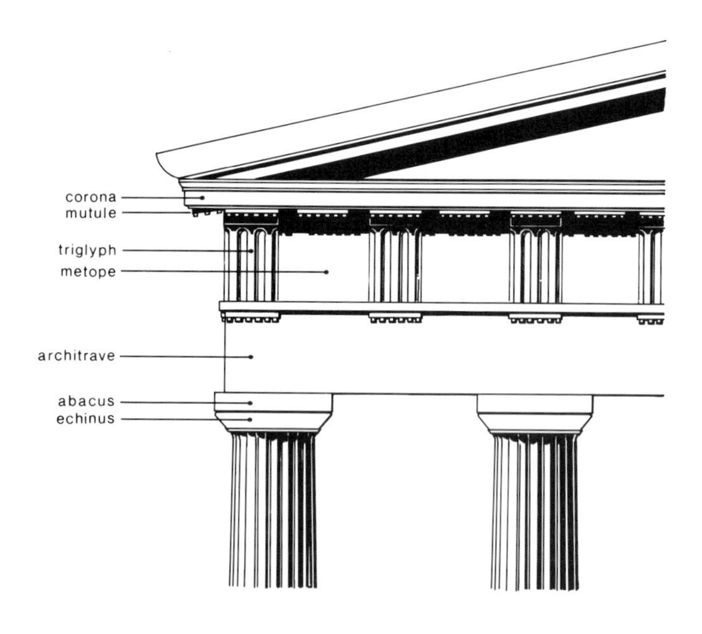

3.25 *Doric temple nomenclature.*

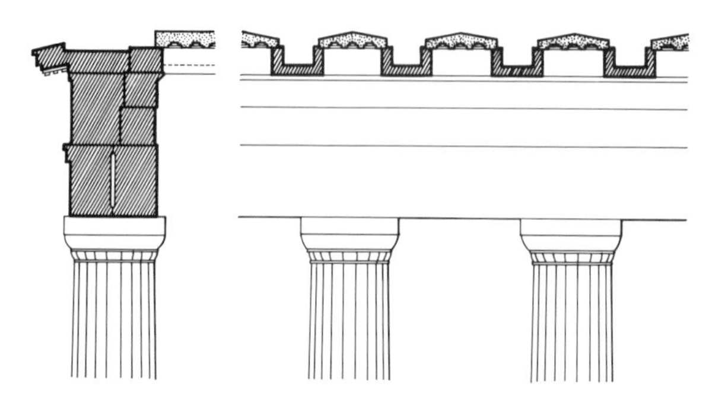

3.26 *Temple of Apollo, Bassai, ca. 400 B.C. Ceiling detail showing hollowed-out marble beams.*

because only the less critical compression side of the beam was reduced; the beam section remains undiminished on the critical tension side. Indeed, the fact that both beams and ashlar blocks of stone were hollowed out in this way suggests that the architects came to realize that the stone in temple walls is substantially understressed in compression.

A surprising technique using iron to enhance the reliability of stone structural members is found at the Propylaea in Athens, built between ca. 437 and 432 B.C. by the architect Mnesikles. The marble ceiling of the Propylaea is supported by an array of beams that in turn rested on Ionic architraves, as illustrated in figure 3.27. Those beams coinciding with columns below the architrave merely transmit forces from above to below, engendering compression; but those over the midspan of the architrave produce significant bending and associated tension (as in figure 3.8b). To reduce this bending—by transferring loading away from the center of the architraves toward the supporting columns—iron bars were set into the top faces of the architraves, with a 2.5 cm-deep gap cut below the bars to allow their centers to deflect under load without coming into contact with the architrave (Coulton, 148–149). In effect, the iron bars acted as independent "relieving beams." Even so, the bars should not be interpreted as behaving in any way similar to reinforcing steel in modern concrete. Modern steel reinforcement does not undergo significant bending; rather the reinforcement functions by accepting direct tension in the beam, thereby relieving the concrete itself from having to resist this pernicious force.

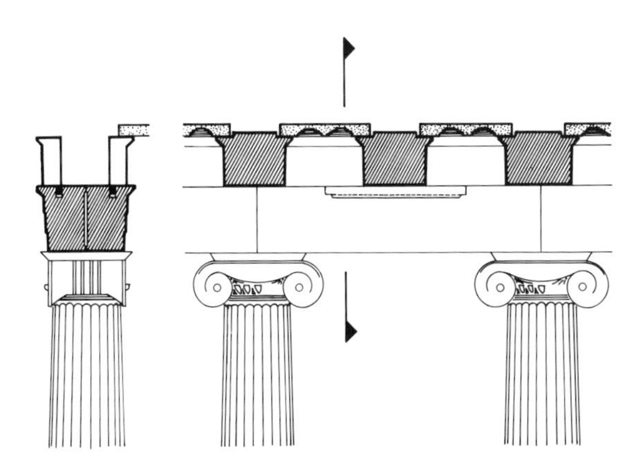

3.27 *Propylea, Athens, ca. 432–437 B.C. Iron-bar insert in ceiling beam (after Coulton).*

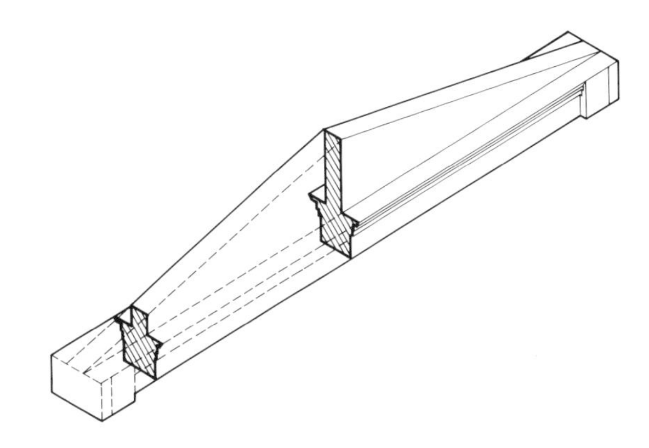

3.28 *Hieron, Samothrace, late fourth century B.C. Deepened ceiling beam (after Coulton).*

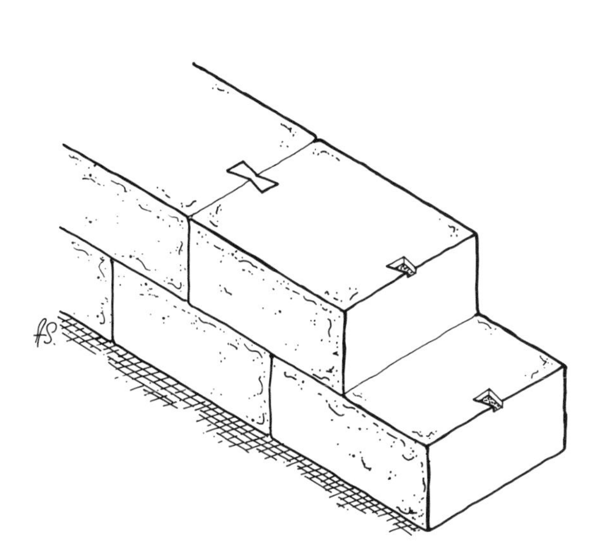

3.29 *"Bow-tie" cramps employed in Greeks temple construction (after Coulton).*

Another structural device, which we have already encountered in the Lion Gate at Mycenae, was used for support of the coffered ceiling of the late-fourth-century Hieron at Samothrace. The ceiling rests on beams that span more than six meters. This considerable span is achieved by having the invisible top faces of the beams extend upward to form a tapered rib (figure 3.28), effectively deepening the beams at their centers.

In lieu of mortar, iron **cramps and dowels** were generally used to fasten together the blocks of stone in Greek monumental construction. Perhaps this use contributed to the Greek practice of building with large stones, effectively reducing the number of cramps needed as well as the total area of carefully dressed stone facing, to eliminate the danger of stress concentrations from rough protrusions. A typical cramp was a bow-tie-shaped iron piece that fit into channels carved into the ends of the stone blocks, stitching them together (figure 3.29). Similarly, iron dowels were used to fasten the drums of columns, preventing them from sliding under shear. The chan-

3.30 *Parthenon, Athens, 447–432 B.C. West Facade.*

nels and holes for both cramps and dowels were cut larger than the iron fittings, so that molten lead poured around the fitting would seal it firmly and help prevent air and moisture from rusting and expanding the metal.

The practice of using such cramps and dowels to resist sliding might seem surprisingly conservative, since the frictional forces between the blocks of stone resulting from the deadweight of the wall would already have been substantial. More likely, the cramps were placed in an attempt to guard against potential stone shifts due to foundation settlements, or possibly, earthquakes. Ironically, the presence of metals probably caused more harm than good: once the buildings were abandoned, scavengers searching for lead and iron gouged holes into the walls of many Greek buildings, weakening them and contributing to their collapse.

Constructed between 447 and 432 B.C. by the architects Iktinos and Kallikrates, the Parthenon has been long noted for its visual refinements and the blending of Doric and Ionic elements (figure 3.30). Its also marks a structural transformation in the organization of the interior of the traditional Doric temple. Coulton has delineated this change by comparing the cross-section of the Parthenon with that of the Temple of Zeus, Olympia (figure 3.31), an important and prestigious mainland temple begun in ca. 465 B.C. (Coulton, 113–117). Although the two temples are of virtually the same height, the Parthenon is wider by more than three meters. This dissimilarity in width reflects differences in the interior span of their cellas: that of the Parthenon is close to ten meters, significantly wider than any other roofed, Doric Greek temple, including the Temple of Zeus whose cella span was less than 6½ meters. Responding to the greater width, Iktinos and Kallikrates altered the six-column portico pattern of the Temple of Zeus to an eight-column pattern for the Parthenon. In both temples, the inner colonnades of the cella are aligned with exterior columns, but at Olympia, the alignment was made with the two central columns, while at the Parthenon, the alignment

3.31 *Comparative sections: Temple of Zeus, Olympia, 470–457 B.C. (above) and Parthenon (after Coulton).*

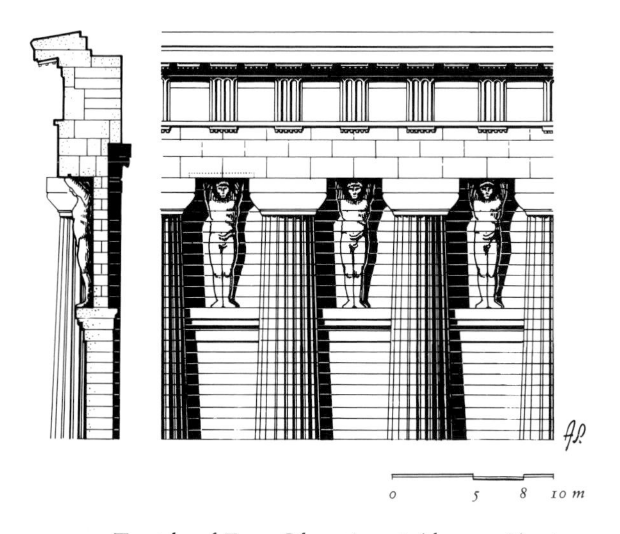

3.32 *Temple of Zeus Olympios at Akragas (Agrigento), ca. 500–460 B.C. Facade with male figures; note iron reinforcement above figure (after Coulton).*

is with the second columns out from the center. This modification not only allowed more space for the cult statue, but it also solved the problem of a large roof span by providing support at shorter, more nearly equal intervals (see chapter 5). Despite the larger scale of the Parthenon, the intercolumniation remains at just over two meters, not unlike that of the earlier temples.

More radical structural innovation was pursued at the giant temple of Zeus Olympios at Akragas (present-day Agrigento) where the great scale of the temple required modification of the basic Greek temple format. Instead of using the more-or-less standard intercolumniation of between two and three meters, the span of the lintels at Akragas would have needed to be extended to a full four meters. The architect compensated for this large span by inserting great columns carved as standing male figures that appear to help support stone lintels at their center (figure 3.32). In fact, the architrave at Akragas does not act as a beam, since the spaces between the columns and the Atlantis statues are filled with coursed masonry walls. But even with these walls, there must have been some lingering concern on the part of the architect for the overhanging portion of the architraves. As shown in section in figure 3.32, iron bars were also positioned in these regions (as in the Propylaea, discussed above). In sum, the classic structural system was eschewed at Zeus Olympios. Only the basic exterior form of the Greek temple was retained and adapted to the new and larger scale.

Beams of marble or stone, carrying the combined loads of entablature and roof, could never be counted on to safely span great distances. The inherent weakness of stone in tension, more than any other single attribute of the classic design, effectively limited the range of variants that Greek architects could devise for the temple form. The limitation was overcome only when a new aesthetic for monumental architecture was introduced into the Roman world.

Imperial Rome

Even before the advent of imperial Roman architecture in the first century A.D., the use of voussoir arches—known but rarely exploited visibly in Greek architecture—came into its own. Roman walls tend to be massive and marked by deep recesses. Primary building materials were brick, stone, and by the turn of the first century for monumental Roman buildings, pozzolan concrete.

Rome and its surrounding region were not favored, as was Athens, with nearby marble quarries. Commonly available stone included peperino, travertine, and a soft volcanic tufa that conveniently hardened when exposed to the atmosphere (Sear, 73). In the provinces, brick and local stone, ranging from the light-colored, fine-grained limestone of southern France to the rough, dark-gray native stone of the Val d'Aosta, were used. Pozzolan concrete, used commonly during the Imperial era, "cures," or sets, chemically in a similar manner to modern Portland cement (see chapter 1 on building materials). This allowed it to be used for underwater construction at harbor sites as well as for constructing aqueducts and sewers, without having the mortar dissolved by effluent. Indeed, this property has contributed to the preservation of many Roman ruins. Even so, Roman concrete differs from modern concrete in that the consistency of modern concrete mixes, composed of water, Portland cement, sand, and fine rock aggregates, is fluid and homogeneous, allowing it to be poured into forms. In contrast, Roman concrete was hand-layered together with chunks of aggregate that often consisted of rubble from earlier buildings. Modern concrete also gains great tensile strength from integral reinforcing steel, whereas Roman concrete, dependent only upon weak cement bonding in tension, could not be relied upon to accept appreciable tensile forces. Because it shared the same weakness under tension, Roman construction in concrete did not really differ from construction in stone or brick masonry. Indeed, the use of pozzolan concrete seems to have been motivated primarily by its economic advantages in construction (Mark 1990, 72). By the beginning of the second century A.D., pozzolan concrete had become the material of choice for large-scale building.

Unlike modern practice, which employs temporary wooden or metal forms to support poured concrete until it hardens, Roman walls, as well as piers and columns, most often used permanent forms of brick or stone, classified according to the pattern of facing used. The three main facing types are: *opus incertum,* an irregular facing of variously shaped small stones, *opus reticulatum,* square stones set diagonally, and *opus testaceum,* coursed brick or tile facing. As illustrated in figure 3.33, facing stones and bricks were usually triangular, serving to increase the surface area between the facings and the concrete core. Found almost universally in early concrete construction, the use of *opus incertum* declined in the last quarter of the second century B.C., though it continued to be used for precinct walls and for rough construction. In finer work, it gave way to *opus reticulatum,* that provided a new standardization of stone blocks, as well as the greater possibility for effects of polychromy, particularly popular until the second century A.D.

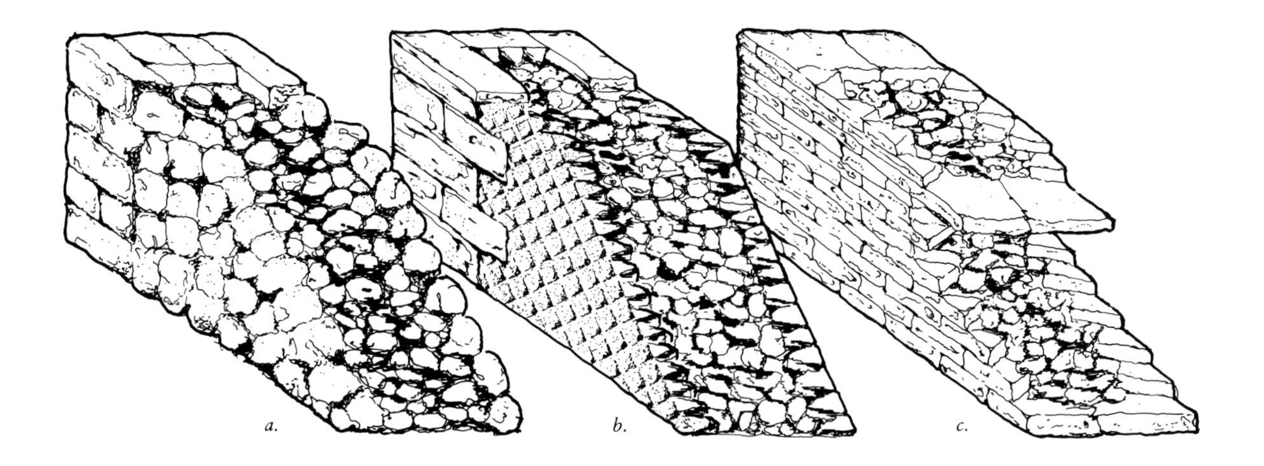

3.33 *Roman concrete facing: (a)* opus incertum*; (b)* opus reticulatum*; and (c)* opus testaceum *(note use of "leveling course" in* opus testaceum *construction).*

As early as the second century B.C., baked bricks and tiles were occasionally used as building material, for example in the Basilica of Pompeii (ca. 120 B.C.). Their application became increasingly frequent in the first century B.C., but it was not until the reign of Nero that *opus reticulatum* was effectively supplanted by *opus testaceum.* **Brick facing** was at first a characteristic of buildings in and near the city of Rome, but the technique was quickly disseminated to the rest of the empire. Brick was especially favored by the Romans because it could be produced quickly and cheaply under industrial conditions, yielding building units of standard size and shape. Roman brick was not configured like its modern equivalent; rather, it had the shape of large, square tiles, typically about 6 cm thick. It was produced in standard sizes up to about 60 cm on a side. When cut along their diagonals, the larger-sized bricks, known as bipedales (two Roman feet), produced four triangular facing bricks. Interestingly, typical Roman mortar-joint thickness did not remain constant over time; the standard, 3-cm joint thickness of first-century brickwork in the capital increased to about 4 cm by the beginning of the second century; by the beginning of the fourth century, it had escalated to some 6 cm, the thickness of the brick itself (Dodge, 112).

Roman concrete walls also incorporated so-called **leveling courses:** through-the-wall horizontal layers, usually composed of large bipedales placed at vertical intervals of about ½-meter. These strata (illustrated in figure 3.33), not keyed into the concrete core above or below, created planes of "potential cleavage . . . a source of weakness" (Ward-Perkins, 99) that could have been troublesome, were it not for the great mass of the wall itself. In addition to providing a level surface at the end of each stage of wall construction, inserts of this type (and similarly, of brick "ribs" in domes) may have played two principal structural roles: the first, to act as a cover for the recently laid concrete during construction, helping to keep it from drying out too rapidly and malforming; the second, to provide deliberate "planes of weakness" (similar to modern expansion joints) that could accommodate structural deformations associated with powerful, thermally induced forces (see Mark 1990, 66) and thereby avert unplanned cracking in regions of the wall remote from the inserted brick layer.

The **amphitheater** was one of the most important public buildings of a Roman city. From the first century B.C., Roman architects were able to reproduce in built form the geometry and substantial seating capacities of "natural" amphitheaters carved of bedrock. In the amphitheaters at Arles, Nîmes, Rome, and Verona, masonry walls converging toward the interior and separated by vaulted passageways provided a surface for the seating ramps. Passages also opened onto circulation corridors from which further slopes or stairs could be accessed. This ingenious interconnection of concentric corridors and inclines facilitated the handling of large audiences. The passageways not only tapered and sloped in a conical form, but were placed at varying angles because the arenas had more than one geometric center. The concentric corridors also required a complex system of vaulting; but our interest here lies primarily in the large curving walls that support the vaults and seating.

The **Colosseum** in Rome, the largest of Roman amphitheaters, measures 118 × 156 × 48½ meters high and provided seating for up to 55,000 spectators. Begun under Vespasian in 75 A.D., the amphitheater was placed on the site of the lake of Nero's Golden House (see chapter 2), thereby transforming part of a palace site into public space. Indeed, the weight of the lake water probably helped to preconsolidate the subsoil prior to the construction of this enormously heavy structure in a low, marshy area (Sear, 135).

Of all the imperial amphitheaters, only the Colosseum possessed two passageways encircling the converging ramps of seats on the same level (figure 2.13). This second concentric passageway permitted the addition of extra banks of seats above, in turn making a third order of arches necessary in the outer wall. Domitian added a fourth story in A.D. 81–82, but all trace of the upper seats has been lost, suggesting that they were constructed of wood. Canvas awnings, known to have been stretched across the arena by a squad of sailors to shield the audience from the sun, are recalled by the stone corbels that once held their supporting masts.

The materials needed to build this massive arena included 100,000 cubic meters of travertine and 300 tons of iron cramps to hold the blocks together (Cozzo, 29–30). Most of the Colosseum was originally covered with a veneer of travertine, scavenged for reuse in the fifteenth century. Too big to be easily carted off, the marble ashlar blocks of the perimeter wall remain, as do brick facings of pozzolan concrete-filled walls. In some cases, exposed pozzolan concrete surfaces remain with the imprint of the formwork still clearly visible. Construction of the Colosseum was completed over such a short period of time that it has been suggested that conventional methods of building (story upon completed story) would have been too slow (Sear, 139). Rather, it appears that construction began with the raising of the outer travertine walls up to the top of the first order, together with the two concentric walls behind them. Piers were then constructed up to the point of the vaults, leaving only a concealed springing for the later vault construction. Finally, brick arches were built to support the sloping barrel vaults, which in turn supported the banks of seats. Virtually the entire skeleton of the Colosseum could in this way have been erected in a short time, permitting large gangs of skilled and unskilled workers to fill in the spaces between the piers later on.

Another aspect of the rapid construction concerns the behavior of fresh masonry in cold weather. Freezing temperatures may damage uncured mortar because expanding ice can literally pull mortar apart. While this problem is endemic to lime mortars, pozzolan concrete cures exothermically; in other words, it produces heat as it combines chem-

3.34 *Colosseum, Rome: buttressing of severed wall.*

ically with water. When poured in large masses, even in cooler temperatures, modern concrete becomes warm to the touch. Roman pozzolana exhibited a similar reaction, and this probably allowed builders to continue construction right through the relatively mild Roman winters.

The extremely tall perimeter wall of the Colosseum benefits from secure bracing at each level of the three superimposed sets of concentric passageways, so that it acts not as a single 48½ meter-high wall, but rather like four individual walls, each one only a quarter of the total height. The upper wall proves least stable, not only because its greater elevation subjects it to higher wind pressure (see figure 3.4), but also because it receives bracing only along its lower perimeter; the other three tiers are securely held in place, at both top and bottom, by the internal vaults. The curve of the arena also contributes a small measure of additional stiffness to the upper wall, but the tremendous weight of the large travertine blocks comprising the wall itself plays a far more significant role in ensuring its stability. As discussed in the introduction, the horizontal thrust of each arch in the arcade is effectively countered by adjacent arches; and because of the closed, oval form, no additional supports were needed to terminate the arcade. (Because of the curvature, the opposing thrusts of the arcades are not exactly aligned, but the resultant small outward forces are easily countered by the building's mass.) After the removal of a portion of the outer wall of the Colosseum, however, a large inclined buttress had to be placed to secure the arcade arches (figure 3.34).

Although used extensively in buildings to introduce large openings for light and access, the voussoir arch is perhaps most clearly studied in the great **aqueducts** erected by imperial Roman engineers. One of the best examples is the 275-meter long Pont du Gard near Nîmes, dating from the early first century A.D. Built to carry water across the valley of the river Gard, the bridge is effectively a wall that rises 49 meters above the river surface, with large openings to permit river flow through the base and to reduce the total volume of quarried masonry on the upper levels.

The construction of the Pont du Gard can be described in terms of a successful formula: large cut-stone blocks laid without mortar to eliminate the danger of water leaching mortar out from the joints and possibly leading to collapse; projecting stones at the springing to provide a ledge on which the flying centering was supported during construction (visible in figure 3.18) and similar stones on the outer faces of the piers and spandrels helped support the scaffolding on which workmen stood and the materials were raised; and the extensive reuse of centering as described above.

Its great width (originally 6½ meters at the lower arcade, 4½ at the intermediate, and 3 meters for the uppermost arcade) and deadweight of its piers and arches more than adequately counter potential bending stresses from lateral wind loadings (as described in figure 3.9). The tiered arches are stabilized by stone surcharges, as discussed in the introduction, through which horizontal thrust is transmitted from one arch to the next, with a zero-resultant net lateral thrust on the supporting piers. At the ends of the arcades, stone cliffs resist the reactions of the outer arches, creating, as the nearly 2,000-year endurance of the Pont du Gard demonstrates, an extremely stable structure.

The Roman Pantheon, built between ca. 118 and 128, presents a different problem in wall structure and design. A massive circular drum wall, composed of a pozzolan concrete core faced with brick and marble, supports the colossal 43-meter diameter dome above (figure 3.35). Although 6 meters thick, the drum is not actually solid throughout. Rather, it consists of eight great piers (shown in plan

3.35 *Pantheon, Rome, ca. A.D. 118–128: interior.*

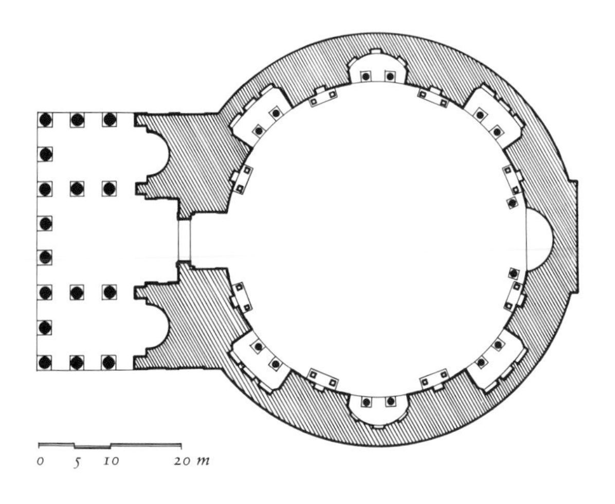

3.36 *Pantheon: plan (after MacDonald).*

in figure 3.36) joined by relieving arches, as illustrated in figure 3.37. Eight deep niches, one forming the entrance and seven others that sheltered statues of the seven major Roman gods, helped to lighten the massive wall and lessen the total load on the foundations. The multiple tiers of arches joining the piers further demonstrate that Roman architects wisely distrusted trabeated support systems for large loads. Even where the decorative articulation on the Pantheon interior appears to be trabeated, relieving arches hidden behind the marble veneer of the walls show the building's dependence on arcuated structure.

Because it had been thought that the step rings surrounding the dome acted to reinforce it and thereby remove much of the horizontal thrust that would otherwise be transferred from the dome base against the upper supporting wall, the thickness of the Pantheon's cylindrical wall drum has been deemed excessive (see, for example, Middleton, I: 66, II: 131, and Robertson, 233–234). However, recent studies have demonstrated the way in which the dome actually behaves as an array of arches that exert immense horizontal forces on the drum wall (see chapter 4). No doubt from earlier experience with arch behavior (as well as with smaller, similar domes), the designer of the Pantheon understood that the supporting structure would need to resist these forces against bending, and he wisely specified its great thickness—in effect, making his wall perform in a manner similar to that illustrated in figure 3.9. The long life of the Pantheon owes much to this crucial design decision.

Like the Colosseum, the Markets of Trajan, constructed in Rome between A.D. 100 and 112, demonstrate the readiness of the Roman emperors to build structures used by the community at large.

3.37 *Pantheon: relieving arches in the wall (MacDonald).*

3.38 *Markets of Trajan, Rome,* A.D. *100–112 "Flying buttresses" of the market hall.*

3.39 *Basilica Nova (Basilica of Maxientius and Constantine), Rome, begun 307: interior reconstruction (after Ward-Perkins).*

3.40 *Basilica Nova: walls of the north exedra.*

The markets housed more than 150 shops and offices in a large complex similar to a modern mall. To buttress the high groin vaults of the main arcade (see chapter 4), the designer employed external braces supported by masonry arches (figure 3.38) not unlike the flying buttresses created a millennium later by French Gothic masons. This innovative structure, however, does not appear to have greatly influenced subsequent Roman builders.

A similar need to provide support for high groin vaults, but at a substantially larger scale than at the Market, was encountered in the construction of the late Roman baths and basilicas. Of these, perhaps the greatest achievement was the imperial basilica begun in Rome by the emperor Maxentius in 307 and completed by Constantine following his victory over Maxentius in 312. The so-called Basilica Nova enclosed more than 1½ acres divided into three high, groin-vaulted bays, flanked on both sides by lower barrel-vaulted bays (figure 3.39). To counteract the thrust of the huge groin vaults, massive lateral buttressing walls, some 4 meters thick, were constructed between the lower flanking bays (figure 3.40). These structures then emerge above what in effect are the side aisle roofs to help support the clerestory walls (figures 3.41 and 3.42). One may also perceive in these projections, with their prominent arched openings, a proto-flying buttress; but structurally, the projections read as more solid spur walls, providing lateral support to the great central vaults. Vertical support for the high vaults was at least partially provided by the great engaged columns (shown in figure 3.40), which effectively reduce the

3.41 *Basilica Nova: fragment structural support for high vault above the north exedra.*

3.42 *Basilica Nova: exterior reconstruction (after Durm).*

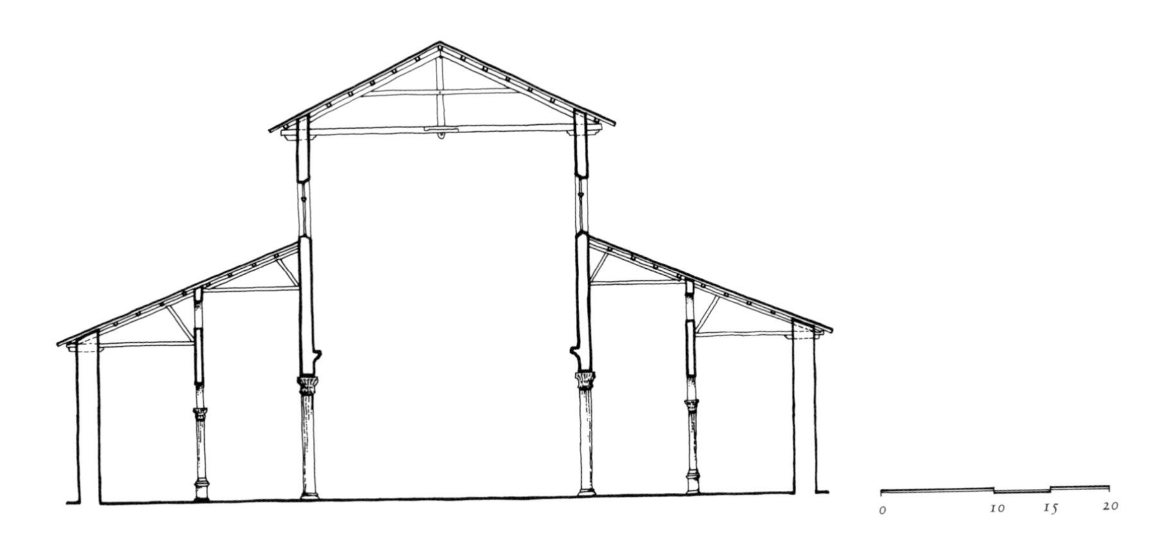

3.43 *St. Peter's Basilica, begun ca. 333: cross-section through the nave (after Letarouilly).*

clear span of the vaults, while the flanking barrel vaults were supported by the massive lateral walls. It is worth noting too that because neither the high groined vaults nor the lower barrel vaults of the basilica receive lateral support from adjacent vaulting at the building's ends, the end walls were built appreciably thicker than the inner ones.

In contrast to monumental imperial vaulted halls, St. Peter's Basilica, like other early Christian basilican churches sponsored by the emperor Constantine, was timber-roofed (figure 3.43). This choice was no doubt influenced by the relative speed afforded by such construction, and it may also have been related to the channeling of vast economic resources toward the construction of a new imperial capital at Constantinople. Yet the more economical timber roof construction was hardly less monumental in scale: St. Peter's boasted a nave clear span of 23 meters! By eliminating the great weight and associated outward thrust components of heavy masonry vaulting (see chapter 4), the walls of timber-roofed basilicas could be much reduced in thickness. As long as the lengthy bottom chords of the timber trusses retained their integrity, wind, seismic loads, and foundation movement would be the only possible sources of lateral forces acting on the walls.

Like earlier imperial basilicas, St. Peter's displayed long parallel walls without buttresses; intersecting cross-walls provided bracing only at the facade and in the region of the transept. The nave clerestory walls were supported on colonnades while the walls between the inner and outer aisles rose above arcades. Lateral stability, therefore, would have needed to be maintained mainly by the exterior walls, with the roof trusses over the side aisles helping to secure the raised clerestory walls to these more firmly rooted structures. By the fifteenth century, it was observed that the southern clerestory walls were inclined as much as two meters out of plumb in certain locations (Alberti, I. 10.). Since the width of these walls could be no greater than 2½ meters (see chapter 2), the resultant force from the deadweight of the wall itself was well outside the wall "middle third" (see figure 3.10). Hence Alberti's inference that the collapse of the nave was prevented only by the propping up of the roof trusses was probably correct; but his placing the blame for the wall distortion on the prevailing northeast wind is less so. As discussed in chapter 2, substantial settlement of the southern foundations under the nave had taken place by that time, providing a far more likely explanation for the existence of the gross wall deformation. Nor would it have been surprising if the tension ties of the roof trusses, the most vulnerable element of a timber truss, had parted due to rot or insect damage by the fifteenth century, placing still additional lateral loads on the top of the wall.

Dating from the same decades as the Basilica Nova, the so-called Temple of Minerva Medica in Rome, also known as the Pavilion in the Licinian Gardens, succeeded in incorporating generous window openings at the base of a large dome. The supporting structure for the dome actually has the form of a decagon, 25 meters across its flat sides, that was composed of ten piers, visually "tied" at several levels by arches. Large, tall windows were set high in the wall between the piers (figure 3.44) above nine projecting apses and the entrance. Apparently the piers exhibited structural distress early on because they were reinforced by increasing their depth (dimension *t* in figure 3.8b), even before the building was completed. Despite the structural problems of the pavilion, however, the pursuit of domed buildings with extensive, high fenestration would continue to occupy the heirs of the Roman designers, as developments in Constantinople were to demonstrate.

Byzantine

The transfer of the imperial capital from Rome to Constantinople in the fourth century meant that there were no nearby sources of pozzolan. Builders in the new capital were thus forced to adapt techniques of Roman pozzolan construction to local building materials. One example can be seen in the massive fortification walls of the city (figure 3.45). Begun in the fifth century but rebuilt and enlarged periodically, they are composed of a mortar and rubble core faced with alternating bands of brickwork and stone (a treatment that continued to be used well into the Renaissance era throughout much of the Mediterranean region).

Walls of monumental Byzantine buildings were normally composed of alternating layers of brick and mortar, with the mortar consisting of brick fragments and dust, including bits of charcoal, that impart semi-hydraulic properties (i.e., the ability to cure chemically and develop good strength—but at a slower rate than mortars based on true hydraulic cements) to its limestone sand mortar base (Penelis, 136–139). Extremely thick mortar joints, often exceeding the thickness of the bricks themselves, today display a hard monolithic mass similar to Roman concrete. Such wide mortar beds would have allowed very rapid bricklaying, perhaps even approaching the speed possible with Roman concrete construction, but their uneven appearance, compared with tightly pointed brickwork, called for a covering with thin veneer. Almost all Byzantine walls displayed stucco, marble veneer, or decorative mosaics.

3.44 *"Temple of Minerva Medica" (Pavilion in the Licinian Gardens), Rome, ca. 320: wall detail.*

3.45 *Constantinople, fifth-century city wall.*

3.46 *Hagia Sophia, Constantinople: leaning south gallery pier with characteristic capital.*

3.47 *SS. Sergius and Bacchus, Constantinople, ca. 527: interior.*

Whereas Roman walls into which niches have been cut appear massive and monumental, most Byzantine interior walls, even when load-bearing, seem visually thin and planar thanks to the veined marble veneer and gold mosaic coverings. The Byzantine sensibility modified even the architectonic forms of the Greek orders, with column capitals drilled to produce fragile, lacy forms that seem to provide little means of supporting the loads above (figure 3.46).

In some ways, the church of SS. Sergius and Bacchus, begun ca. 527 in Constantinople, provides a transition (if at a much reduced scale) between the fourth-century Roman Pavilion in the Licinian Gardens and Hagia Sophia in Constantinople, undertaken some five years later. Recalling the pavilion, support for the sixteen-sided "pumpkin dome" with windows at its base is provided by eight (rather than ten) piers that form an octagon 16 meters across the flat sides (figure 3.47). Unlike the pavilion, however, the church is largely rectangular in plan except for its apse (figure 3.48), resembling more closely the exterior form of Hagia Sophia. Its exterior walls do not reflect the internal buttressing providing lateral support to the piers under the dome.

Justinian's great domed church of Hagia Sophia (Holy Wisdom), constructed in Constantinople between the years 532 and 537 (figure 3.49), also looked back to the Roman tradition, particularly the Basilica Nova, the Pavilion in the Licinian Gardens, and the Pantheon. Its innovative design, by the architects Anthemius of Tralles and Isidorus of Miletus, combined the traditional longitudinal basilican plan with an enormous central dome. By combining the two building forms, the architects succeeded in fusing ecclesiastical and imperial liturgies.

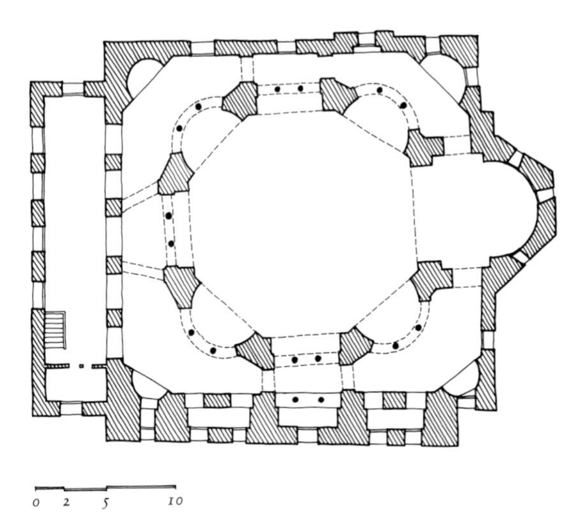

3.48 *SS. Sergius and Bacchus, ground plan.*

Given the close correspondence in scale between the original dome of Hagia Sophia and that of the Pantheon (see chapter 4), it is likely that the Pantheon provided the principal structural model for Justinian and his architects as they translated Roman concrete into Byzantine brick. Although the behavioral implications of this translation from concrete to brick are less significant than might be supposed, because the walls and dome of the Pantheon are unreinforced and have little strength in tension, Hagia Sophia was a precedent-setting building in many other ways. Where the vast dome of the Pantheon rests on continuous massive, niched walls, four great arches and a like number of pendentives direct the weight of Hagia Sophia's superstructure to four prodigious supporting piers (see figure 3.1) so the tympanum walls need merely to sustain their own deadweight. The tympani are pierced with many windows (whose original size was larger than the present openings), bringing light directly into the central space of the church. The combination of such large glazed surfaces with a dome of monumental

3.49 *Hagia Sophia, Constantinople: air view.*

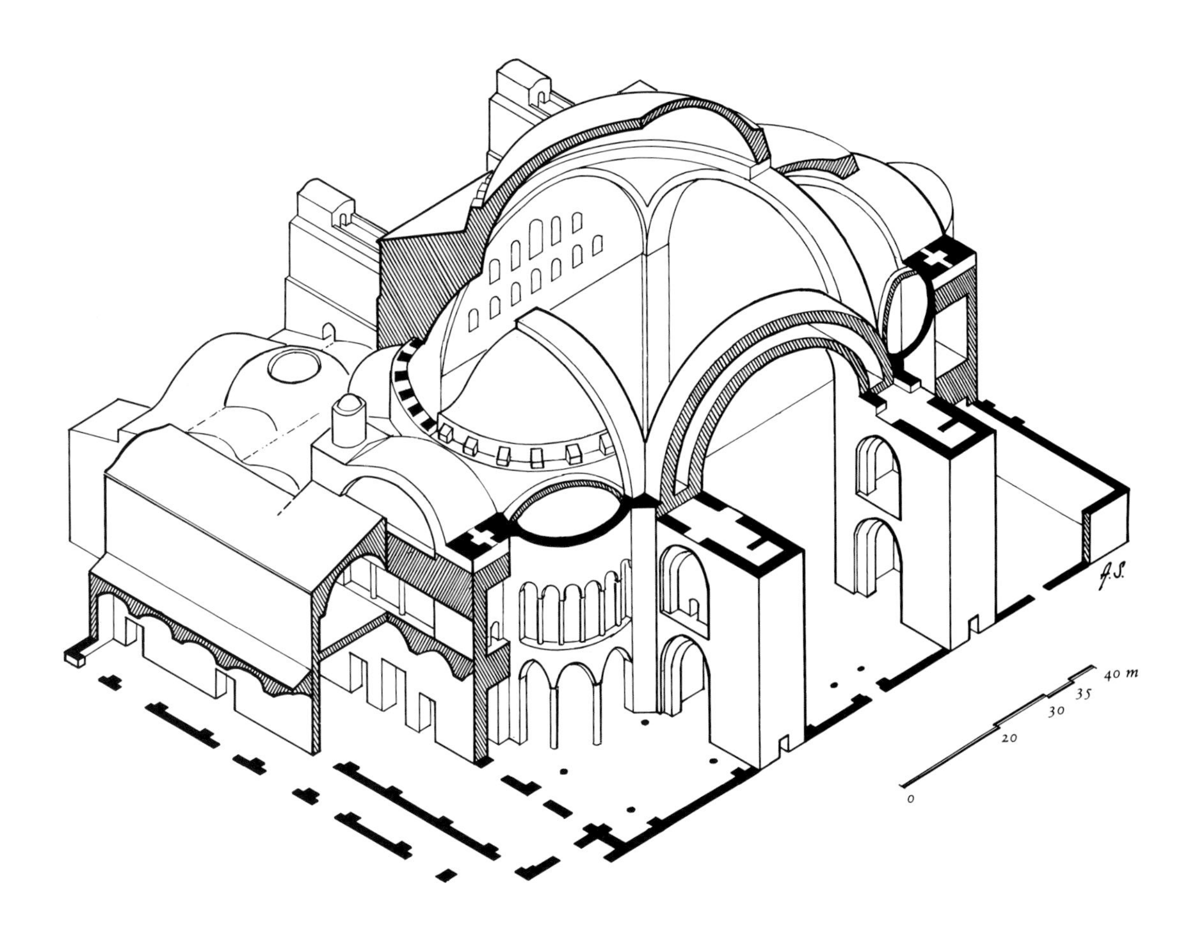

3.50 *Hagia Sophia: analytical drawing of the structure.*

scale had never before been attempted and stands as one of the greatest architectural achievements of the age.

Compounding the physical and structural challenges of the design was political instability within the empire that required Hagia Sophia, as *the* most visible symbol of Justinian's power and prestige in the capital, to be completed as swiftly as possible. Construction proceeded in more or less horizontal layers until the erection, in ca. mid-535, of the main arches to support the dome. These are 31 meters in span and spring some 25 meters above the floor (figure 3.50). "Flying centering" of the type illustrated in figure 3.17 was probably used for assembly of these arches. In all likelihood, the centering would not have been adequately tied to prevent immense horizontal forces from impinging on the upper portions of the main piers, which then proceeded to tilt outward. Though they are essentially secondary structure, the aisle and gallery colonnades also display the effects of the deformations. As illustrated in figure 3.46, their outward inclination is startling.

Historians have pointed to the speed of Hagia Sophia's construction as contributing to the great distortions, reasoning that lime mortar would not have had sufficient time to set in the massive piers before they were loaded laterally by the great arches. Yet the main piers are not of brick and mortar, but rather of cut stone, at least up to the level of the gallery floor. It is, rather, the four great pier buttresses intended to brace the main piers to the north and south that are constructed of brick with mortar joints appreciably thicker than the individual bricks. (The main piers, because of their form with greater depth in the east-west than in the north-south

IN TEMPLO

directions, as well as the bracing eventually received from the east and west semidomes, are effectively more rigid in the east-west direction than to the north-south.) The problems of the buttresses became evident to the builders, and before proceeding further with the erection of the main dome, the exterior pier buttresses were reinforced and raised to their present height (see figure 3.50). Raising the buttresses and thereby increasing their weight provided additional stability by effectively reducing the ratio of bending to direct loading (in the manner described in figure 3.9). The buttresses must have then seemed secure, because the dome was raised in time to allow the vast building project to be completed in five years.

The great central dome fell in 558 after being subjected to two major earthquakes: one in August 553, and the second in December 557. A second dome, having a higher profile than its predecessor, was then erected in 558–562. Despite partial collapses after an earthquake in the tenth century, and again after another in the fourteenth, the general form of the second dome today remains unchanged from that of 562. But structural repairs associated with these incidents, as well as other adversities, involved the placement of much additional buttressing around the entire structure.

The Justinianic buildings of Ravenna serve to bridge the gap between the Byzantine and western medieval traditions. The most famous of these, the church of San Vitale, resembles SS. Sergius and Bacchus in its general form (figure 3.51). San Vitale, however, took advantage of a western European technique for lightweight vaulting (see chapter 4).

3.52 *Old cathedral (the "Basse Oeuvre"), Beauvais, eleventh century: west face.*

3.51 *San Vitale, Ravenna, ca. 530–548: interior.*

Thus freed from the need to buttress a heavy dome, the upper walls were pierced with large clerestory windows, while the lower levels of San Vitale shimmer with mosaics. Yet despite its adopting western-style construction, both the mosaic surfaces and the use of long thin bricks at San Vitale follow the style of Constantinople.

Early Medieval

Next to the massive scale and advanced technology of the Byzantine achievement at Hagia Sophia, developments in the treatment of walls during the early medieval period in the West seem comparatively insignificant. Several notable innovations, however, were combined with the well-preserved Roman architectural heritage. Rubble core walls were generally used with facings composed of stonework, reused ashlar, pillaged brick, or stucco. In the absence of pozzolan concrete, the rubble cores employed lime mortars of varying qualities. Few early medieval walls remain that have not been continuously maintained to prevent rainwater from washing out those cores.

Decorative applications analogous to those used in Roman practice were common, but in at least some early medieval buildings, this decorative appliqué also formed an important element in the wall structure. The eleventh-century Basse-Oeuvre in Beauvais, for example, preserves much of its original decorative facing of octagonal stones. These stones, which appear to be planar facing, actually penetrate 25 cm into the rubble core, forming a thick stone formwork, much like Roman *opus reticulatum* (figure 3.52). Similar octagonal blocks penetrate the entire thickness of the walls of the Carolingian Lorsch Abbey gatehouse forming not only a striking decorative surface, but the structural core of the wall as well (figure 3.53).

3.53 *Abbey Gate House, Lorsch, late eighth century.*

Classicized wall construction was commissioned by ambitious dynasties attempting to invoke imperial Roman tradition. Although Merovingian and Ottonian rulers also adopted this strategy, Charlemagne himself made the most direct architectural allusion to antiquity in his Palatine Chapel at Aachen (figure 3.54) which frankly recalls, among other ancient and classical monuments, Justinian's San Vitale. Begun in 792 under the direction of Odo of Metz, the character of the Carolingian building differs radically from that of the prototype. To a large extent, these aesthetic differences stem from the choice of materials. The walls of Charlemagne's

3.54 *Palatine Chapel, Aachen, begun 792: elevation reconstruction (after Schneider).*

church were executed in cut stone rather than the mosaic-encrusted brick seen at Ravenna. Moreover, the heavy stone dome at Aachen requires thick, small-windowed walls for support. These walls are further buttressed by stone galleries about the perimeter of the central domed space.

In part, the use of stone as the principal building material in western early medieval architecture reflects the availability of good building stone in much of northern Europe as well as economic and demographic changes in the post-Roman world, which supported labor-intensive stone cutting over the industrial production methods of the empire. Although the aesthetics and structural behavior of Charlemagne's chapel remain similar to Roman antecedents, the shift in the north away from the predominant brick and rubble and mortar construction to construction with large cut stones marks a decisive change in primary construction of the monumental buildings of this region for the balance of the Middle Ages.

Romanesque

The Romanesque period ushered in a new era of complexity in wall structure. Increasingly taller Romanesque walls supported massive stone vaults as well as heavy timber roofs. It is important to recognize, however, that during this period walls also became more complex even when the spaces they framed were unvaulted.

Coined in the nineteenth century, the term Romanesque itself arose in recognition of the similarity to Roman forms. Ordinarily constructed of rubble cores with brick or ashlar facings, related to the earlier Roman *opera,* Romanesque walls tended to be relatively massive, layered, and articulated with round-headed openings at all levels, much like Roman precedents.

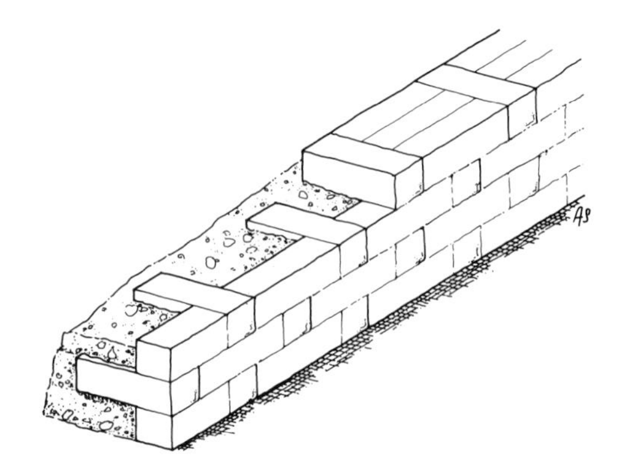

3.55 *Romanesque header-stretcher stone-laying technique (after Adam).*

Fortuitous natural conditions near a building site sometimes led to building techniques such as "integral coursing," like that found at Aulnay (figure 3.21), where each course can be followed around the entire building, across openings, inside and out. In other instances, which may derive from Carolingian precedent, the outer facings are built integrally into the rubble core, as in the so-called *opus monspeliensis* at Maguelone Cathedral and other Romanesque buildings in southern France. At Maguelone, a uniform ashlar block is laid in a header-stretcher fashion (figure 3.55) so that a positive tie is created with the rubble core, and at the same time a decorative surface is achieved with the alternation of their long and short sides. In regions where geology did not permit such stonework, masonry coursing was less regular. In some major buildings, such as St. Sernin in Toulouse and St. Alban's in England, reused Roman brick replaced stone as the primary building material.

3.56 *St. Mark's Venice, 1063–ca. 1095.*

Romanesque builders had to make do without the fine pozzolan mortars found in Roman building. Higher-strength mortars from natural cement sources were available in a few localities, as at the trass deposits on the Rhine River, or the "scaly, rugged" stone mentioned by Palladio near Padua (Palladio, I. 5.), but such deposits were probably uncommon and did not influence the main line of architectural development until well into the Renaissance.

Although rubble core walls bound with relatively weak lime mortars performed well when initially built, water inevitably seeps through them over time and tends to leach out the mortar, leaving substantial voids that allow the two outer faces to separate and bulge outward and, sometimes, even to collapse. Although this phenomenon probably did not develop as a major problem during the first several generations after construction, structural problems and continual repair are typical for Romanesque buildings, as the example of the eleventh-century Church of St. Mark's in Venice (figure 3.56) demonstrates. Although adopting the five-domed Greek cross layout from the (long-ago destroyed) Justinianic church of the Holy Apostles in Constantinople, the solidity of the walls at St. Mark's does not compare with that of typical Justinianic walls, and this has led to continuing structural and maintenance difficulties (Krautheimer, 41). A comparison of Romanesque walls with their Roman antecedents shows how the use of pozzolan mortars result in far better preservation despite their greater antiquity.

The tall walls of Romanesque churches sometimes stemmed from a giant-order nave arcade, as in the Auvergnat churches like Notre-Dame-du-Port in Clermont-Ferrand or in west French hall churches like St-Savin-sur-Gartempe near Poitiers. It

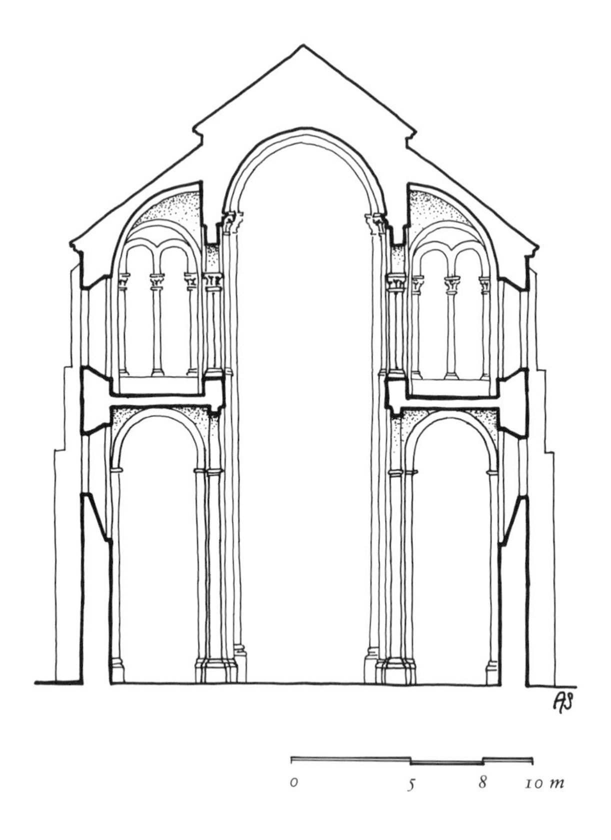

3.57 *St. Foy, Conques, begun ca. 1050: cross-section of the nave (after Mark).*

3.58 *St. Foy, Conques: nave interior.*

also resulted from the superposition of bands of arcades as at St. Sernin in Toulouse or St. Alban's in England. In either case, lateral stability of the high nave wall was assured, following Roman precedent, by flanking, vaulted side aisles and, often, vaulted galleries, as illustrated by the cross-section (figure 3.57) of St. Foy at Conques (ca. 1050–ca. 1125). This scheme was adopted in all the so-called major pilgrimage road churches that included, in addition to St. Foy and St. Sernin, Santiago de Compostela in Spain and two others now destroyed, St. Martial at Limoges, and St. Martin at Tours. The galleries in all of these buildings functioned liturgically as well as structurally, since they could be used for circulation by pilgrims. But because the galleries permitted only indirect lighting of the main vessel, the high vaults and the upper regions of the nave walls were left in shadow (figure 3.58).

In other buildings like the unvaulted abbey church of Jumièges, near Rouen, and the third church at the abbey of Cluny, the clerestory walls were opened to permit light to enter the upper zone of the main vessel. At Jumièges (ca. 1067) the walls, reaching up 23 meters, are comparatively flat and appear to be quite thin (figure 3.59). To sustain lateral loadings from wind and the weight of the roof, the tall, thin walls were thickened appreciably just below the clerestory windows and were buttressed from behind by vaulted tribunes built above the side aisles. The walls were also reinforced on the exterior by wall buttresses and on the interior by applied colonnettes on alternating piers. It is probable too that the nave was divided by diaphragm arches crossing the major (reinforced) piers. Diaphragm arches act like transverse walls, helping to stabilize the thin nave wall along its more than 40-meter length as well as to provide additional support to the roof.

Where masonry vaults rise above the clerestory, as in the now largely destroyed, giant third abbey church at Cluny (1088–ca. 1121), the high walls, in addition to supporting wind loadings and

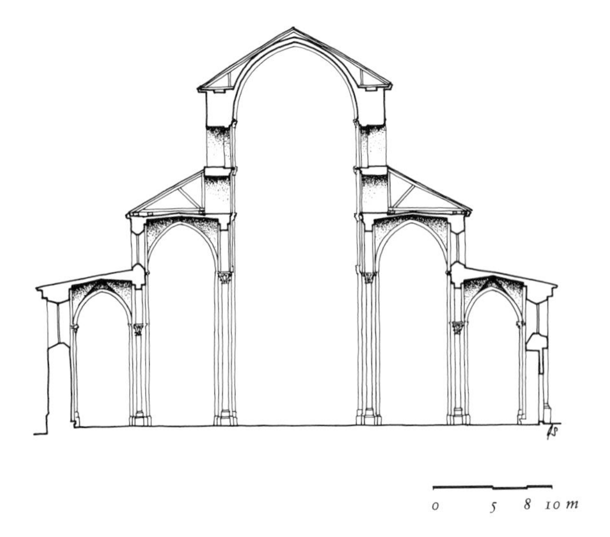

3.59 (Opposite page) *Abbey Church, Jumièges, north interior wall of the nave.*

3.60 (Left) *Third abbey church, Cluny, 1088–ca. 1121: interior reconstruction (Conant).*

3.61 (Above) *Third abbey church, Cluny: reconstruction of the cross-section through the nave (after Conant).*

the weight of the roof, must also resist the substantial outward thrusts of the vaults (figure 3.60). Wall buttresses, usually coinciding with the transverse arches on the underside of the high barrel vaults, help to stabilize such walls at regular intervals. Cluny and its early-twelfth-century cousins at Autun and Paray-le-Monial, relied mainly on wall buttresses for the security of their clerestory walls. But because of the magnitude of the bending forces in the pierced upper walls, there were inevitable design limitations. Windows needed to be small and few in number. The clerestory of Autun is pierced by only one window, which occupies less than a third of the bay wall length. At Paray-le-Monial, there are three windows per bay, but the height of these is but a small fraction of the height of the full clerestory. The almost 27-meter-tall clerestory walls of Cluny, pierced by three more generous windows per bay (figure 3.61), collapsed within five years after their erection. Little is known of the disaster, which might have been caused by weakness of the walls or by problems with the foundations (see chapter 2). It is known, though, that the clerestory walls were quite heavy, almost 2½-meters thick, and that lightweight materials had been used for the high vaulting to reduce thrust (Conant, 203).

Consequently, in barrel-vaulted churches, even with clerestories, the lighting of their central vessel tended to be diffused and limited. Only in special regions, as at an octagonal crossing tower or in the curving wall of an apse which was usually substantially lower than the central vessel, could the walls be opened with larger windows. In these regions, the curving form of the wall itself (tending to develop a more effective cross-section in plan than that of a linear wall for limiting stress and deformation) serves to stiffen the wall and allow for penetration. Intense direct light in Romanesque churches was intermittent and related to function. The nave and choir, when long enough, were comparatively dark and punctuated by intervals of more intense lighting, which signaled the liturgically important areas of the crossing bay and the apse, the usual loci of the two main altars.

With the new height of buildings in the eleventh century, upper chapels and rooms also became more common. Mural passages and internal stairs, located most often in crossing towers and in west facades, began to be used to gain access to these upper areas. The passages and galleries not only facilitated circulation; they also provided a structural response to the problem of stiffening taller walls. Comparing the nave wall of Jumièges with that of the originally unvaulted contemporary nave of the abbey church of Saint Etienne at Caen (figure 3.62), we can appreciate this innovation, the so-called *mur épais* (or "thick wall," by Bony 1939). The *total* thickness of the "walls-with-voids" at St. Etienne exceeds that of the high walls at Jumièges, but consist of thinner wall sections joined by narrow barrel-vaulted passages. As discussed in the introduction, walls depend primarily on their thickness for lateral stability, so that a wider wall overturns far less easily. In essence the *mur épais* accomplishes the effect of increasing the width of the wall without requiring a corresponding increase in the amount of quarried and transported stone. But because the two walls tend to shear, or slide relative to one another when subjected to lateral force, to be effective the connections between the inner and outer wall sections must be secure, constraining them to act as a single unit. If the barrel vaults of the passages are well placed to resist these shearing forces, the *mur épais* will be far stiffer under lateral loading than a conventional wall

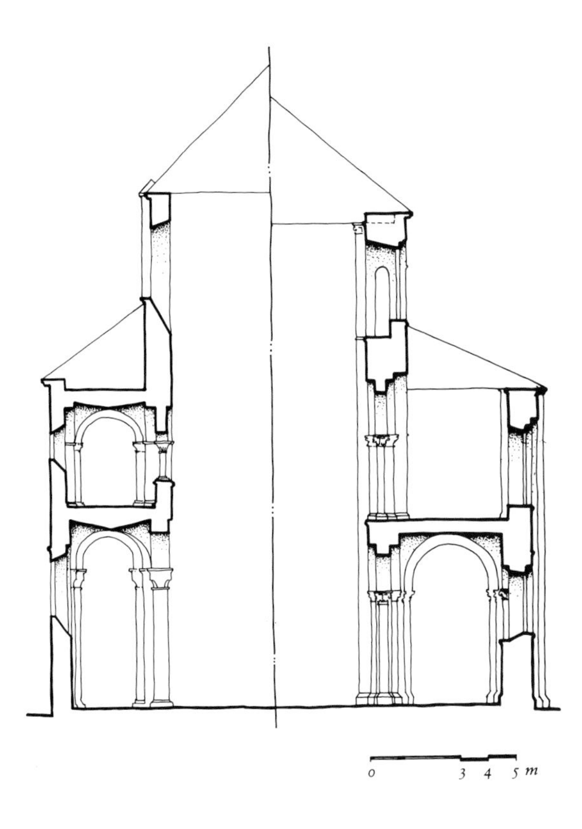

3.62 *Comparative nave sections: (right) St. Étienne, Caen, begun ca. 1065, and (left) Abbey Church, Jumièges (modified after Dehio and Bezold).*

of equal weight. In this regard, it may be well to add that the original nonthrusting timber-trussed roof over the nave of St. Etienne was replaced by ribbed vaults in ca. 1130–1135; and although wall motion has brought some distress to the vaults over the years, it does not appear to be very serious (Mark 1982, 112).

While it could be difficult to argue that the widely dispersed Romanesque building projects prefigured the development of the Gothic style, the era certainly did signal an increased level of structural and stylistic experimentation that, with hindsight, does indeed point to many of the events that followed.

Gothic

Structural, spatial, and aesthetic innovation evolved together to an extraordinary degree in Gothic architecture, spawning unprecedented wall configurations having greatly enlarged fenestration set within increasingly light and tall structural systems. Although much of the impetus for the new Gothic style came from incorporating ribbed cross-vaulting (discussed in chapter 4) that focused supporting forces at discrete points on the wall and thus fostered a more skeletal system of support, the Gothic adoption of the planar, pointed arch was equally crucial. The advantage of the **pointed arch** is more spatial than structural. Like its semicircular predecessor, it is composed of circular arcs that could be easily laid out by the early builders (see chapter 1 on design). But unlike the semicircular arch, whose fixed center demanded that its rise be always half the span, the location of the centers of the pointed arch segments was fluid (see figure 3.63). The flexibility of the

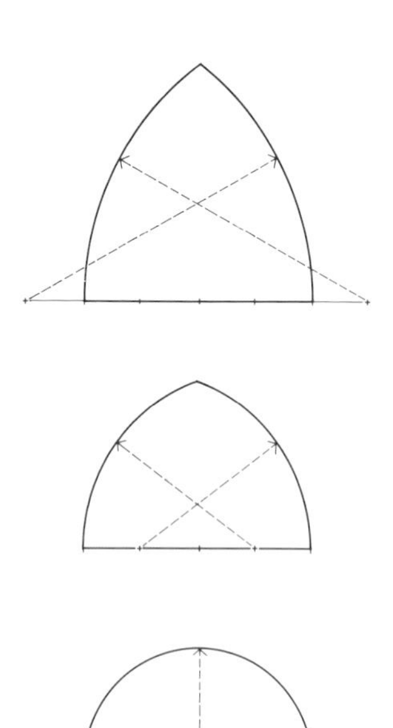

3.63 *Laying-out a Gothic pointed arch (after Viollet-le-Duc).*

3.64 *Timber props supporting the weakened wall of a building in Leon.*

pointed arch allowed its adaptation to almost any spatial need. There were structural benefits as well, but these have been exaggerated by commentators who have not taken into account that the longer, pointed arch employs more stone than a semicircular arch of the same span, and hence that the weight of the pointed arch is greater. But it does benefit from the fact that the horizontal component of thrust of any arch against its abutments varies inversely with its rise. And since pointed arches rise higher than semicircular arches, they tend to generate less thrust, even if this reduction is somewhat offset by their greater weight.

The **flying buttress**, the third principal attribute of Gothic structure, relates to both the adoption of the groined vault and the pointed arch. It functions simply as a linear brace—in the manner of the timber prop illustrated in figure 3.64—to resist the focused thrust of a vault or of wind loading on a great roof. The brace itself is normally composed of one or more rows of ashlar masonry all acting in compression, which in turn are usually supported from below by a segmental arch (figure 3.7). The effect of all of these devices was to pare down the supporting structure of Gothic walls, leaving the interstices almost entirely free to be opened up to fields of brilliant stained glass—the aftermath best judged by a comparison of the interior walls of the choir of the early-twelfth-century giant Romanesque abbey of Cluny (figure 3.61) with the remodeled abbey church of St. Denis (figure 3.65) realized a little over a century later. Below Cluny III's pointed barrel vault were relatively small windows that emphasized characteristic Romanesque wall thickness and mass. On

3.65 *St. Denis, choir and northeast transept, ca. 1240: wall elevation; cf. figure 3.60.*

the other hand, the tall, open clerestory and triforium set below the high quadripartite ribbed vaults of St. Denis imparts a sensation of weightlessness.

While there were earlier buildings of the Ile-de-France that anticipated later Gothic development, the birth of the Gothic style is generally dated to 1144, when Abbot Suger consecrated the new choir at St. Denis. Although the relatively small spans and only moderate elevation of the twelfth-century choir do not display anything like the structural pyrotechnics of later Gothic structure (for example, the later thirteenth-century building at St. Denis itself), the use of supple, ribbed vaulting and large stained-glass windows set within a highly voided non-load-bearing wall was (and still is) most dramatic (figure 3.66).

The Cathedral of Sens, contemporary with St. Denis but planned on a much larger scale, was begun in the Romanesque style and then altered to bring it more up-to-date. Notwithstanding its Gothic elements (figure 3.67), it retains the Romanesque elevation that typically comprised no more than two or three stories. During the second half of the twelfth century, masons of the Ile-de-France vigorously pursued development of wall design. This so-called early Gothic period saw an increasing emphasis on height, generally achieved through the use of four-story elevations. In addition to the three stories found at Sens, the great cathedrals of the early Gothic period incorporated fully vaulted galleries. These buildings, largely constructed between 1160 and 1190, include the cathedrals of Laon (figure 3.68) and Noyon, and the delicate south transept of Soissons Cathedral.

Begun around 1155, the early Gothic cathedral of Notre Dame in Paris was planned to be the tallest building in France. Compared to its next-tallest Gothic predecessors, Sens and Laon, its nave

3.66 *St. Denis ambulatory, 1140–1144.*

3.67 *Sens Cathedral, north wall of the nave, ca. 1145–1164.*

was one-third again as high, although its construction was generally lighter. With a keystone height at 33 meters and a steeper timber roof than found in earlier buildings, the wind pressures impinging against its upper walls and roof are significantly greater than in the lower buildings with which its designers would have had experience (see figure 3.4). The combination of higher pressures with the presentation of larger areas of resistance to wind called for a new and radical solution. But unfortunately, substantial reconstructions made to Notre-Dame after 1225 have obscured exactly how this design problem was originally solved.

From archaeological clues in the structure of Notre Dame and in contemporaneous buildings in the Paris region, as well as from drawings and photographs made before another major campaign of restoration in the mid-nineteenth-century, a new reconstruction of the original structural configuration of the nave was determined that incorporated probably the first medieval use of flying buttresses (Clark and Mark, 51). The horizontal thrust caused by the dead loading of the nave vaults was mainly resisted by the masonry arches that also supported the roof above the gallery (figure 3.69). Whereas the walls of earlier churches had been adequately braced against live wind loadings by the structure supporting the vault, the considerable wind loadings on the unprecedented tall clerestory and roof at Paris could no longer be effectively countered at this level. Perhaps because of observing cracking during construction, or possibly, movement of the clerestory and parapet walls, flying buttresses were added to increase the building's resistance to wind.

Model analysis (figures 3.70 and 3.71) of Notre Dame's reconstructed original nave structure suggests that local cracking would still have been

3.68 (Opposite page) *Laon Cathedral, ca. 1170: north wall of the nave.*

3.69 (Above) *Paris, Notre Dame Cathedral, reconstruction of the nave, ca. 1180 (Clark and Mark).*

3.70 Photoelastic modeling.
Analysis of the long vessels of Gothic churches is facilitated by their repeating modular bay design, as illustrated in figure 3.69. The structural support of each nave bay is provided by a series of parallel transverse frames comprising piers, buttresses, lateral walls, and ribbed vaults. Moreover, structural forces within the masonry are hypothesized to be distributed as they would be in a homogeneous, monolithic material; but for this to be so, it must also be assumed that the entire "frame" is in compression. This assumption coincides with criteria for successful long-term masonry performance because medieval mortars cannot withstand tensile stresses over long periods of time (see chapter 1, Building Materials). Previous studies indicate that compression does indeed prevail throughout Gothic buildings, and that regions of tension, where they do exist, are highly localized (Mark 1982, 21). A final major modeling assumption is that the heavily loaded foundations give complete fixity to the bases of the piers and buttress walls.

In the modeling process, stress-free dimensionally similar (to the prototype) models of epoxy plastic are loaded by scaled arrays of weights representing the distributions of wind and deadweight on the actual structure (see figure 1.4). Test are performed in a controlled-temperature environment where the loaded model is first heated to 140° C. and then slowly cooled to room temperature. Model deformations at the higher temperature are "locked in" after cooling so the loadings can be removed with negligible effect. The deformed model, now viewed through polarizing filters in the illustrated polariscope, displays patterns which, with calibration and scaling theory (see Mark 1982, 24–26), can predict the force distributions in a full-scale structure.

3.71 *Photoelastic interference pattern in windward buttressing of the reconstructed structure of the nave of Notre-Dame de Paris under simulated wind loading. Arrows designate regions of local tension.*

detected following heavy storms, such as modern records indicate could have taken place from time to time during the forty-odd-year life of the original structure. The tensile cracks would have required continual, expensive maintenance, which probably led to the decision to rebuild the cathedral structure. But the experience was not a complete loss; its meaning was applied by masons to the taller cathedrals begun at Bourges and Chartres near the close of the twelfth century, and then brought around full circle to the thirteenth-century rebuilding of Notre Dame itself.

The flying buttress was quickly recognized as an important device to liberate tall Gothic walls from the earlier constraint of stacked galleries. Among the first buildings to take full advantage of the new technology was the cathedral of Bourges (figure 3.7), a remarkably efficient and elegant structure. A close relative to Notre Dame, Bourges incorporates light, steeply sloped flying buttresses which, recent studies have revealed, were apparently superimposed on an initial flying buttress configuration much like the conjectured original design of Notre Dame (Mark and Clark, 181). Similarly, on the interior Bourges adopts much of the Parisian spatial scheme except for the omission of transepts and another, more crucial change: the gallery floors have been eliminated, creating extraordinarily tall nave arcades and inner aisles. Most notably, this disposition of internal structure allows far more *direct* interior lighting than the configuration of Notre Dame, where the light from the gallery windows has to pass over gallery floors. The direct light from the lower clerestory windows at Bourges proves most effective at floor level since these windows appear less foreshortened than those of the upper clerestory (figure 3.23), as well as their being closer to the floor. The

structure of Bourges is also exceptional, displaying the lowest stress levels found in any of the great Gothic cathedrals, while at the same time using a relatively small quantity of stone for its structure (Mark 1982, 36).

Despite its successes, however, the influence of Bourges on the future course of Gothic design was to be less than that of Chartres Cathedral, which displays far less efficient structure (cf. figures 3.7 and 3.72). Scholars have speculated that this influence may derive from the greater flexibility of the Chartres formal and structural scheme (Branner 1950, 169). In any event, it was Chartres, along with its smaller contemporary, the nave of Soissons, that pointed the way to Gothic architectural development.

The era inaugurated by the work at Chartres, usually referred to as the High Gothic, is generally regarded as the golden age of the Gothic style. The High Gothic elevation typically consisted of a tall arcade and similarly tall clerestory, separated by a narrow, dark triforium. The dark triforium, embodying a stone wall behind a screen of columns, may have been seen as desirable for aesthetic reasons, since it provided a visual balance point between the brightly lit zones above and below. It corresponds to the "attic" below the side aisle roofs, and originally it hid the transverse, triangular spur walls that provide lateral support to the piers as well as to the side aisle roofs at each bay. With the advent of the flying buttress, the heavy spur was no longer needed, but the designer of Chartres still retained it in his design (see figure 3.73; note that the spur had been abandoned at Bourges). In fact, the buttressing of the Chartres nave, which uses the equivalent of three

3.72 *Chartres Cathedral, begun 1194: buttressing of the nave.*

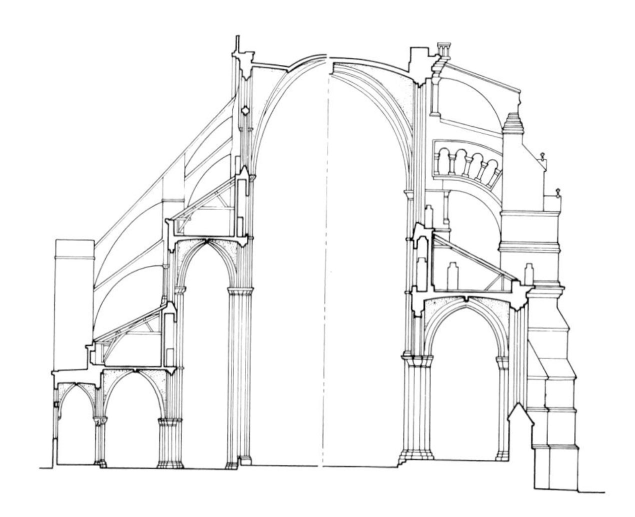

3.73 *Comparative sections: (left) Bourges Cathedral, choir, and (right) Chartres Cathedral: nave (Mark).*

3.74 *Reims Cathedral, begun 1210: cross-section of the nave (Mark).*

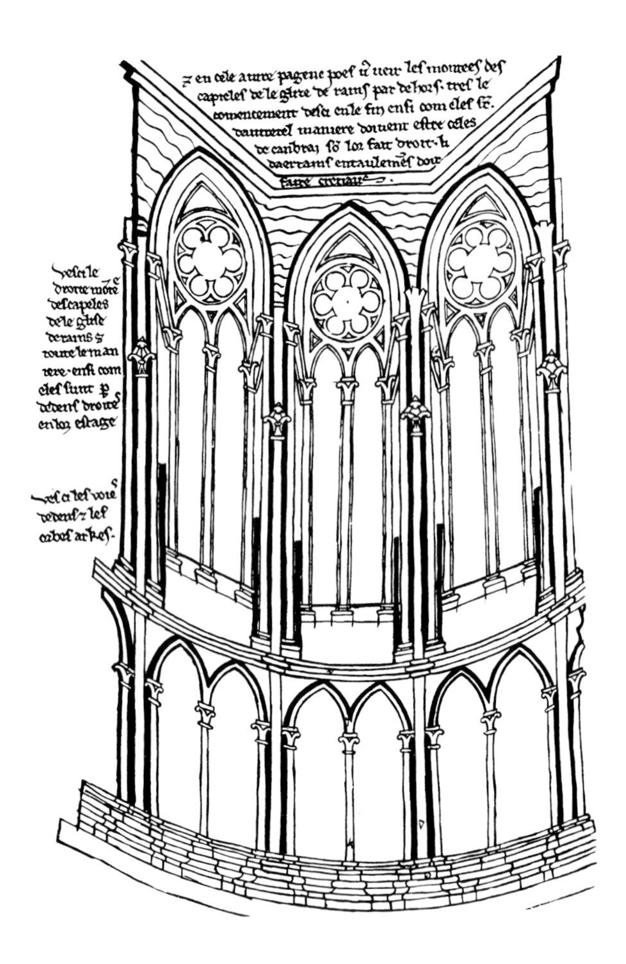

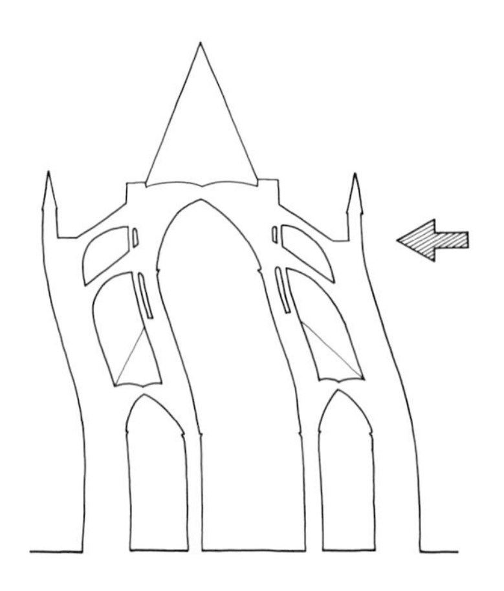

3.75 Reims Cathedral: bar tracery in the upper hemicycle (Villard de Honnecourt).

3.76 Exaggerated deformation of Gothic building frame under the action of lateral, wind loading.

3.77 Amiens Cathedral, begun 1220: flying buttresses and pinnacles atop nave buttress.

separate flyers and the spur wall, has been characterized as being technologically clumsy (Mark 1990, 113). Even so, Chartres seems to have pointed the way to the far more efficient buttressing systems of Reims and Amiens that support their nave walls with but two specialized buttresses, as illustrated in figure 3.74: a lower flyer to receive the outward thrusts of the high vaults, and an upper flyer, strategically placed to receive the thrust of high wind loadings on the tall superstructures.

The cathedral of Reims, begun in 1211, adhered to the High Gothic elevational system of Chartres, generally repeating the nave arcade/triforium/clerestory composition, but on a larger scale. As the coronation church of France, Reims was also executed with lavish decoration and sculptural adornment. In particular, the Reims workshop substituted bar tracery for plate tracery in the upper portions of the clerestory zone. While earlier rosettes consisted of large carved stone frames, the first architect of Reims, probably Jean d'Orbais, redefined these stone surfaces as a series of thinner stone "lines" or bars erected within the window void (figure 3.75). This formal treatment underlines the non-load-bearing nature of the wall, with the windows functioning as curtain walls, hung between the far heavier supporting structures of piers and exterior flyers.

Generally considered as the apogee of French High Gothic achievement, the cathedral of Amiens demonstrates, at least in its 42-meter-tall nave (figure 2.23), keen attention to details of structural performance. In high winds, the walls of framed buildings tend to bend into an S-shape (as illustrated

in figure 3.76) because of the restraint of the ties between the walls, which in a Gothic church consist of the high crossing ribs, vaulting, and the heavy principal framing of its timber roof. This characteristic structural response was found, from model testing, to produce tension in the upper region of the leeward pier buttresses of the Amiens nave during exceptional storms (Mark 1982, 55). Furthermore, it was found that the weight of a pinnacle placed on the outside corner of the pier buttress compensates for this effect (figure 3.77); that is, the pinnacle engenders enough compression just beneath it to overpower the effect of the tension. This positioning of the pinnacle does not contribute to the *overall* stability of the buttress (although the pinnacles are relatively light compared to the great weight of the buttresses, and could not affect stability very much), but must instead have been motivated by observation of tensile cracking in this zone, likely in one of the first bays of the nave during erection. Clearly, the Gothic master mason was intimately involved with the building process and receptive to incorporating design changes in response to on-site empirical observations.

By the mid-thirteenth century, the High Gothic was succeeded by the Rayonnant style, which brought the voiding of the wall to its natural conclusion. The glazing of the triforium, initiated most probably at the Cathedral of Troyes or in the remodeled parts of St. Denis (figure 3.65), and applied in the eastern transept walls and choir of Amiens, eliminated the last blank stone surface in the Gothic wall. Spur walls had already been cast off, and the timber shed roofs that had formerly blocked the light to this zone were replaced with flat or pyramidal roofs. And with this gesture the structure had evolved fully into an armature, where the glass walls could be given over to the display of elaborate tracery; yet an unforseen problem set into motion by the presence of a pyramidal side-aisle roof might have led to the collapse of the tallest of all the Gothic cathedrals.

The original choir of the Cathedral of Beauvais, begun in 1225 and whose vaults reached up to 48 meters (figure 3.78), stood for only twelve years before the vaults collapsed. Too easily interpreted as a Gothic Babel Tower doomed to failure by its audacity, a less simplistic view of Beauvais and its place in history reveals much about the Gothic design process and demonstrates the importance of economic factors in the patronage of large architectural projects.

Model analysis corroborated by archeological investigation pointed to a problem in the Beauvais design at the junction of the intermediate pier buttress with the aisles (figure 3.79). Because of Beauvais's open triforium and the accompanying covering of this crucial region of the structure by a pyramidal timber roof, the masons may have failed to make adequate inspection and take corrective action (Mark 1982, 132). Instead of succumbing simply to fate, then, Beauvais appears to have been a victim of identifiable design flaws which, undetected, proved fatal to its tall, delicate structure.

The notable decline in northern French building after 1284 is often attributed to a lack of confidence in tall-building safety, set off by the Beauvais collapse. In fact, the inactivity was caused mainly by a downturn in the local economy (Strayer, 57–58), and soon thereafter, also by the outbreak of

3.78 *Beauvais Cathedral, begun 1225: choir interior.*

3.79 *Beauvais Cathedral: choir section (Mark).*

the Hundred Years War between England and France. Enormous late Gothic churches, in Metz, Narbonne, Palma (on the island of Majorca; figure 3.80), Milan, Seville, Ulm, and others, scarcely smaller, at Bologna, Gerona, and Prague, attest to the continuing vitality of the Gothic style throughout the rest of Europe (although the southern churches, generally topped by much less steep roofs than those of the north and subjected to gentler winds, did not present similar critical problems of wind loading; Mark 1982, 98). Indeed, the Gothic still flourished in Germany after the fourteenth century when a modified type of basilica-church interior elevation became more common. In these late Gothic so-called **hall churches,** which include the Cathedral of Vienna (ca. 1370–1433) and the Lorenzkirche in Nuremberg (begun 1439), the heights of all three aisles were made almost equal. By eliminating the high central clerestory of the nave, a large floor plan could be created without the expense of carrying the building interior to extreme height. Yet a great timber-framed roof covering all three aisles, as illustrated in figure 3.81, would display a tall exterior. Flying buttresses were replaced by wall buttresses, or simply by reinforcing the relatively low exterior walls by a projecting masonry "leg" between the windows.

More is known about the design techniques used for these buildings than about High Gothic planning, for several of the architects' notebooks from this later period have been preserved (Shelby 1977). We are also in possession of certain structural design rules, as set out in the *Instructions* written in 1516 by a master mason at Heidelberg, Lorenz Lechler, for the benefit of his son (Shelby and Mark). In addition to presenting geometric schemes for planning the configuration of hall churches, Lechler gives specific advice on structural details such as wall

3.80 *Palma Majorca Cathedral, nave interior, begun 1357.*

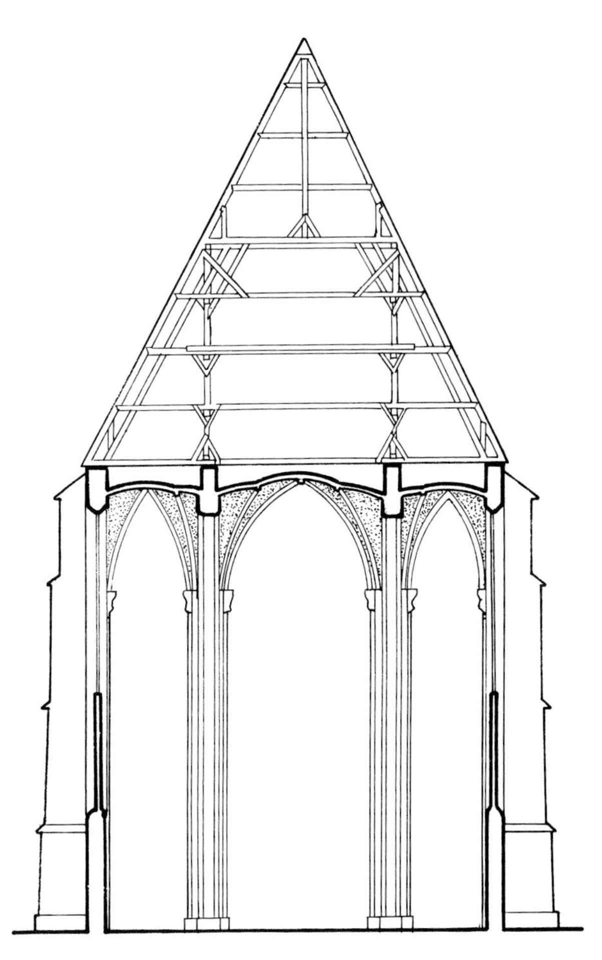

3.81 *Cross section through a typical late Gothic German hall church (Mark).*

thickness, window-opening sizes, and buttress and vault-rib dimensions.

Although Lechler does not clearly state their interdependence, the dimensions of all the building elements are related to the interior span of the central aisle, which he recommends keeping between 6 and 9 meters. For example, he advises that the height of the central-aisle vault keystone be made equal to one and one-half or two times the span of the central aisle. Wall thickness is essentially one-tenth of the span, as is the breadth of the outstanding leg of the wall buttress at ground level. The buttress extends from the outside edge of the wall about two-tenths of the span, giving a total wall/buttress depth at ground level of three-tenths of the span of the central aisle. Thus, for a church with a 9 meter central-aisle span, the vault keystone elevation may be 13½ or 18 meters, while the total depth of buttressing in both cases is about 2⅔ meters. Holding the buttress depth constant for a range of building heights makes sense only with regard to resisting the outward thrust of the vault, where stability against overturning is a function of the total wall/buttress weight. As discussed in figure 3.6, the higher vault would produce greater bending at the buttress base, but this in turn is resisted by the greater weight of a correspondingly higher buttress. Evidently, Lechler's rule does not take into account the effects of wind, but since the buildings with which he is concerned are relatively small, the omission is not serious.

Lechler's criteria seem to have been based on his observations of preexisting buildings. Although he did advise his son to use his own judgment and not necessarily to "follow [the text] in all things," the very existence of the instructions must have had the effect of standardizing building form. With no theory offered to guide experimentation, the only prudent course for assuring safe structural design would have been to follow the text in detail. And as discussed in chapter 6, this would eventually serve to stifle structural innovation.

Although the cathedrals and other major churches of the era provided the most visible and spectacular embodiments of the Gothic achievement, secular architecture also evolved dramatically. Military, civic, and domestic structures all responded to the same economic and political forces as did the Church, and incorporated many formal features of ecclesiastical buildings.

Fortification wall development was marked by continuity with the Romanesque tradition. For obvious reasons, function takes precedence over style in a structure whose form might determine life and death. Fortifications were often larger, better funded, and more widespread than other public construction projects, yet they were often designed, like walls in most structures, to carry the loads of masonry vaults or timber roofs. Others, like the massive precinct walls of the town of Carcassone built in the thirteenth century, were designed to be virtually free-standing structures in themselves (figure 3.82). These were crowned with wall-walks and shooting platforms, and contained gallery passages along their length as well as garrison rooms in the towers. "Live loads" experienced by military walls included troops and munitions, as well as the pressures of assault. The arsenal of attack included a number of features that exploited the potential structural instability of the wall to advantage: battering rams to encourage shearing (figure 3.83), and sappers to tunnel under the wall or, more often, to undermine one of its corners. In turn, military designers answered with thickened walls to protect against shear failure, deeper and thicker foundations to avoid tunnelling,

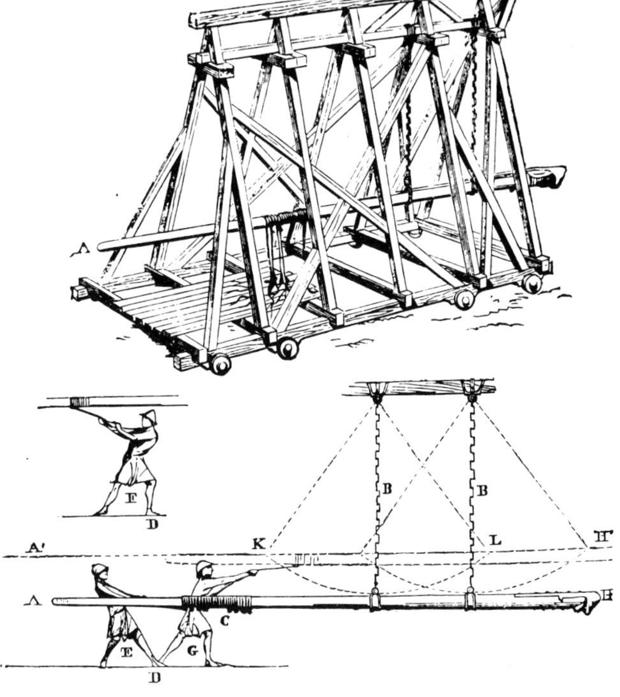

3.82 *Carcassone: thirteenth-century precinct walls (pre-restoration photo).*

3.83 *Medieval battering ram (Viollet-le-Duc).*

and round corners to prevent the prying out of corner stones. Round towers also gave the added advantage of increased stiffness to the wall. The talus, or sloped base (figure 3.84), similarly had a dual advantage. It was developed as a defensive measure against sapping and overturning, but it could also be used as a ricochet to distribute more widely missiles dropped from openings in the wall above.

Fortifications also developed features seldom seen in nonmilitary walls: round profiles and unusual thickness, as well as the additional projections of fighting platforms, garrison stations, machicolations, and crenelations. All these traits were developed primarily as responses to military attack and siege. On the other hand, castles and churches sometimes shared a common vocabulary or construction techniques. As in Romanesque and Gothic church construction, castles witnessed a new and widespread use of wall passages and galleries executed in ashlar. In an ecclesiastical context, these mural passages were meant primarily to lighten the wall and to carry more effectively the loads of heavy timber roofs and stone vaults. In castles, these passages also lighten the upper stories of extremely thick walls. At Dover, for example, the base of the wall is nearly 4 meters thick. If carried solid to roof height, this wall would require massive foundations. Instead, the walls were hollowed with gallery passages. These galleries not only reduced the load of the wall; they also provided a system of communication and protection for the garrison within the skin of the structure.

Civic pride, in the late Middle Ages, also motivated the construction of many large parish churches, town halls, and towers. In Germany, parish churches such as those at Dinklesbuhl and Nordlingen have dramatic towers. Even larger were the ca-

3.84 *Fortified-wall talus (Viollet-le-Duc).*

3.85 *Strasbourg Cathedral: spire, 1439 (1851 photo).*

thedral spires of Vienna, Ulm, and Strasbourg, the latter, at 142 meters, the tallest structure of the Middle Ages (figure 3.85). Other significant monuments of more explicitly civic flavor include the Palais de Justice in Rouen and the town halls of Bruges and Brussels. Indeed, the Gothic era brought new elegance to even utilitarian secular structures. The high quality of workmanship in stone and timber that had been established in the workshops of the great cathedrals carried over into the construction of houses, hospitals, and guild halls. There can be little doubt that at least the upper classes in northern Europe enjoyed, by the fifteenth century, a fairly high standard of living, enriched by the craft-oriented traditions of their culture. But meanwhile in Italy, the seeds were being sown for a revolution in architectural design that would fundamentally alter the relation of craftsmanship and design.

Renaissance

Traditional practice no longer served as the wellspring of architectural knowledge in the Renaissance. It came to be replaced by reliance on the authority of ancient architecture. Filippo Brunelleschi (1377–1446), Leon Battista Alberti (ca. 1404–1472), Donato Bramante (1444–1514), and Andrea Palladio (1511–1592) developed architectural skills not through exclusive training at the building site but rather by the study of ancient ruins. Alberti perhaps best illustrates this new phenomenon: a humanist by training who studied Roman ruins while working as secretary at the Vatican, he later employed craftsmen-architects on site who directed the construction of his designs. Moreover, the "discovery" of the treatise of Vitruvius by Poggio in the St. Gall monastery in 1415 dramatically altered the course of Renaissance architectural ideas. Although the technical content of *De architectura* is not completely valid, it served as the model for architects of the fifteenth and sixteenth centuries to follow in their own writing. This is not to say that Renaissance writers neglected contemporary building practice. In fact, much of the technological exposition in their treatises simply recapitulates the traditional "rules of thumb" preserved previously in the guilds.

Renaissance writers understood enough about stability of walls, for example, to recommend lessening their thickness as they rose, saving the expense of extra material in the upper portions and providing a sufficiently wide base to prevent overturning (see figure 3.6). They also took care to recommend a symmetrical arrangement in freestanding walls and supports: "the wall's thickness should diminish equally on each side of the centerline, columns should be placed exactly above one another, and windows should likewise be placed over one another so as to leave a solid, unbroken vertical support wall" (Palladio, I. 1.).

Architects clearly appreciated that not all sections of a wall support equal loadings. Alberti, for example, divides the wall into the separate parts of *os et complementum,* or bones and paneling. By "bones," Alberti indicates "corners and inherent or additional elements such as piers, columns, and anything else that acts as a column and supports the trusses and roof arches" (Alberti, III. 6.). In this, he is actually referring to wall buttresses; corners are generally reinforced, either by using larger blocks of stone, sometimes accentuated by their projection and/or contrasting color, or by pilasters and engaged columns. He also noted the Roman practice of using periodic leveling courses that "acted as ligatures, or muscles, girding the structure together," and the role of cornices in securing all the walls to each other (Alberti, III. 8.).

Alberti was concerned with the quality of building materials and their application. He recognized the need to select stone for exterior walls based on its capacity to resist weathering as well as strength, and he noted that stone must be oriented with its bedding planes parallel to the ground in order to avoid splitting due to compressive loads. He also understood that mortar does not cement walls together into a monolith, but rather acts primarily to distribute forces evenly across the surfaces of the stones. Such understanding of architectural technology was probably common among master builders of earlier eras, but it was only after Alberti's and later treatises that there was wider dissemination of such knowledge.

Perhaps because of the import attached to the writing down of design rules, structural innovation became less of an issue in the architecture of the Renaissance, an outlook that can be gleaned even from the Renaissance architect's treatment of walls. Unlike the diaphanous northern medieval wall, most Renaissance walls were built in imitation of the massive works found in the ruins of imperial Rome. Niches and applied pilasters were intended to impart a similar monumental quality. Even so, the use of stone-faced rubble core walls tapered off, especially in the more modestly scaled palaces and villas. Part of the reason for this is geographical: the cultural and artistic movement we know as the Renaissance was primarily an Italian phenomenon, and brick was the common building material throughout much of Italy. Unlike ashlar, whose high quarrying and shaping costs proved economically unattractive, brick was relatively inexpensive and was therefore found prudent to use for construction of solid walls. Brick also has the advantage over equally economical irregular shaped stones because of its superior performance under loading. Its parallelepiped form insures the transfer of compressive forces perpendicular to the interface joint, whereas in a strongly compressed wall, weakly bound stones with *inclined* faces, such as those constituting the earlier Roman *opus incertum* and *opus reticulatum,* tend to work their way out of a wall. The Romans could depend on pozzolan mortars to affix such stone, but with less reliable mortars at their disposal, Renaissance builders preferred the predictable squared surfaces of brick.

The few large-scale projects of the era that did require outstanding structural and constructional innovation were great domed buildings, including the cathedrals of Florence and London, and the new basilica of Rome. The earliest of these projects, at Florence, called for an immense dome over earlier Gothic construction. Hence this project is best described in chapter 4; background on the design of the supporting walls for the other two major building projects follows.

The building site for the new basilica of St. Peter's remained active for almost the entire sixteenth century. The first proposal to replace old basilica came in the 1450s during the pontificate of Nicholas V. But although the idea was discussed throughout the second half of the fifteenth century, nothing was actually done until the election of Julius II in 1503. His plans for St. Peter's began with his tomb in a chapel appended to the old Constantinian basilica, but the dilapidated condition of the ancient structure, together with a good deal of ambition on the part of both the pope and his architect, Bramante, encouraged the decision to tear down the old building and replace it with an edifice worthy of the seat of the western church.

3.86 *St. Peter's Basilica, Rome: Bramante's great crossing piers in construction (Van Heemskerck, ca. 1530).*

Given the enormous Pantheon-sized dome they would eventually have to support (see chapter 4), vertical elements were a vital part of the structural system of the new basilica. Yet Serlio accused Bramante of cutting corners and using inferior materials in the construction of the four principal piers that were to support the dome (figure 3.86), in order to complete as much work as possible while the enthusiastic Julius II still lived (Serlio, III. 6. fol. 16). It is likely, however, that Bramante simply underestimated the structural forces accompanying such a monumentally scaled dome. He had never before undertaken a project of this scale, and later project architects, notably the craft-trained Antonio da Sangallo the Younger (ca. 1483–1546), would spend much time and effort increasing the size of the piers.

Owing to the large scale of the building, the walls of St. Peter's are extremely thick. Many passages within the walls provided access for workmen and help lighten them, as with the *mur épais* of the northern Romanesque buildings. In fact, Michelangelo (1475–1564) was able to build a spiral stair within the walls large enough to be used by small draft animals, to aid in carrying building materials for the dome up to the roof. Structurally, the thickness of these walls helps to prevent overturning from the thrusts by the large vaults above the nave and aisles.

Sir Christopher Wren's St. Paul's Cathedral reflects an unusual combination of Gothic building tradition and Renaissance aesthetic. Although Wren (1632–1723) was a scientist by training, he became familiar with medieval building techniques from projects of restoration and while removing the burnt-out remains of the Norman/Gothic St. Paul's after the Great Fire of 1666. So solid were the piers of the old cathedral that Wren resorted to blasting to remove them, at least until he was compelled to stop by nearby property owners.

Wren drew on Roman precedent for the new design of St. Paul's, but even with the backing of Charles II, he could not prevail against the insistence of the high church clergy that the layout of the new building follow "cathedral fashion"—that it have the cruciform, Latin-cross plan of a medieval church. This issue, probably more than any other, required Wren to produce and submit a series of designs and to make major concessions before work on the new St. Paul's finally began in 1675.

In the cathedral as it finally materialized, Wren effectively created two buildings. The interior, modeled on a medieval basilican plan and elevation with a high center aisle flanked by lower side aisles (figure 3.87), even reflects the exterior flying buttresses intended to help support the domical vaults. Nothing of this form, however, is suggested to a viewer on the street to whom the great central dome appears to rise from a massive two-story base (figure 3.88). The cross-section of the nave (figure 3.89) reveals that this effect was achieved by raising the perimeter walls so that they conceal both the inner features of the building and the flying buttresses. The walls, replete with false "windows," are nothing more than "screens" calculated to hide the Gothic-derived clerestory and flying buttresses of which Wren disapproved. Yet the extreme weight of these walls has also given rise to the thought that they were intended to play a structural role, perhaps to resist the thrust of the flying buttresses.

To try to answer the question of Wren's purposes, with an eye toward the possibility that he employed scientific insights, the structure of a typical bay of St. Paul's was analyzed using photoelastic modeling in the same manner as that used to study

3.87 (Opposite page) *St. Paul's Cathedral, London, 1675–1710: interior, looking east.*

3.88 (Above) *St. Paul's, London: view from the southeast.*

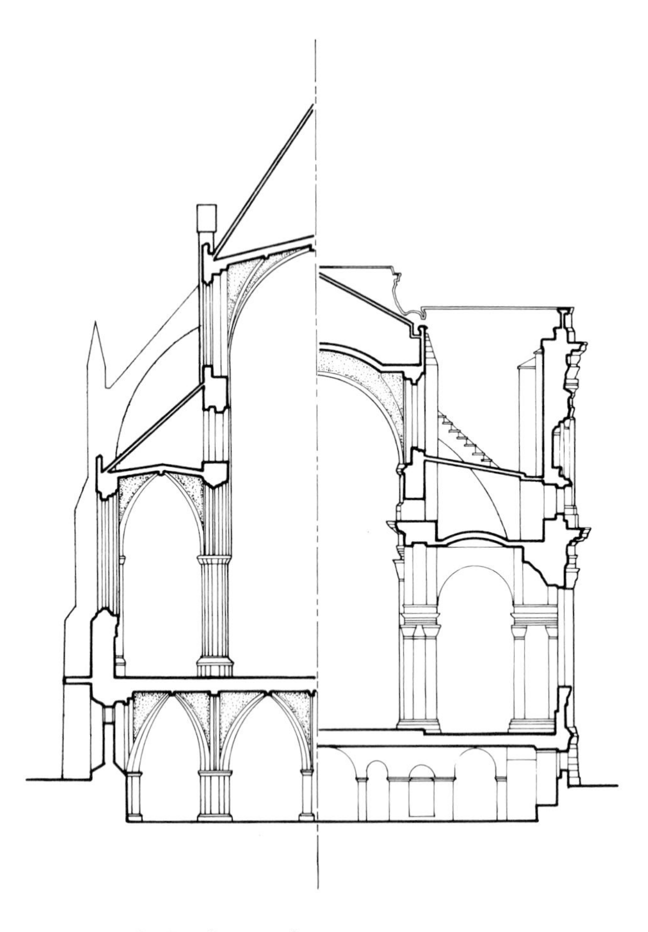

3.89 *St. Paul's, London: comparative cross-sections: left: Gothic choir of the old cathedral; right: Wren design (Dorn and Mark).*

Gothic structures (see figure 3.70). The model was tested first under simulated loadings scaled to actual deadweight, and then a second test of the model was performed by applying scaled wind loadings. After the first series of tests was completed, the flying buttresses were removed from the model and it was tested again under the scaled deadweight and wind loads. Without the flying buttresses, stresses in the piers were higher, but it was ascertained that well-constructed piers of solid masonry (or at least piers having a firm outer shell composed of several layers of coursed masonry) could successfully resist both the thrusts of the vaulting and the effects of high winds. Wren's flying buttresses appear therefore unnecessary.

In view of his earlier warning that because of their "being so much exposed to air and weather, [flying buttresses] are the first thing that occasion the ruin of Cathedrals" (Wren, 298), it seems ironic that they were employed at all in St. Paul's. Of course, Wren may have believed that both the flyers and the heavy perimeter walls were necessary for the building's structural integrity. Yet if he was thinking only of structural necessity, he could have provided discrete piers placed to receive each flying buttress but still hidden behind light walls along the perimeter of the cathedral. The height to which he carried these massive and expensive walls suggests that he was using them primarily for visual effect in the classic tradition.

Even more ironic is the fact that the additional weight of the parapets produces high enough compressive forces to cause spalling in the masonry. The outer stones of the walls of St. Paul's were precisely cut on the outside surface to give only a very thin mortar joint. Because of the expense of finishing the entire surface of a stone block evenly, stonecutters cut back the inside edges so that a precise surface would not be required. The additional space was then filled in with a thick bed of mortar, leaving a relatively crude joint inside but a precisely detailed joint on the outside. The mortar between the blocks has shrunk slightly, causing the outside face of the block to carry a disproportionate portion of the compressive force. The resulting high stresses have caused the stone to spall at the joints, weakening the wall and providing an entry point for damaging moisture. Nor was this just a long-term problem: the building records of St. Paul's indicate similar problems of spalling in rubble-filled piers as early as 1690 (Fürst, 114). It may be well to remark that during the construction of St. Paul's, Wren was concerned also with the progress of a great many other building projects—leading to the observation that problems of this type were more likely to be diagnosed and corrected early on with the more intimate connection between the architect and the building process of earlier eras.

Bibliography

Adam, Jean-Pierre. *La Construction Romaine: Materiaux et Techniques*. Paris, 1984.

Alberti, Leon Battista. *On the Art of Building,* trans. J. Rykwert, N. Leach, and R. Tavenor. Cambridge, MA, 1988.

Bony, Jean. "La technique normande du mur épais à l'époque romane." *Bulletin Monumental,* 96 (1939), pp. 153–188.

Branner, Robert. *The Cathedral of Bourges*. Cambridge, MA, 1989.

Bruno, V. J., ed. *The Parthenon*. New York, 1974.

Clark, William W., and R. Mark. "The First Flying Buttresses: A New Reconstruction of the Nave of Notre-Dame de Paris." *Art Bulletin,* 66 (March 1984), pp. 47–65.

Conant, K. J. *Carolingian and Romanesque Architecture 800–1200*. New York, 1959.

Coulton, J. J. *Greek Architects at Work: Problems of Structure and Design*. London, 1977.

Cozzo, Guiseppe. *Il Colosseo: l'Anfiteatro Flavio nella Techica Edilizia, nella Storia delle Strutture, nel Concetto Esecutivo dei Lavori*. Rome, 1971.

Dehio, Georg Gottfried, and G. von Bezold. *Die Kirchliche Baukunst des Abendlandes,* 7 vols. Stüttgart, 1884–1901.

Dinsmoor, W. B. "The Architecture of the Parthenon." In Bruno, ed., *The Parthenon,* 1974.

Dodge, Hazel. "Brick Construction in Roman Greece and Asia Minor." In S. Thompson, ed., *Roman Architecture in the Greek World*. London, 1987.

Dorn, Harold, and R. Mark. "The Architecture of Christopher Wren." *Scientific American,* 245 (July 1981), pp. 160–173.

Fitchen, John. *The Construction of Gothic Cathedrals*. Chicago, 1961.

Fürst, Viktor. *The Architecture of Christopher Wren*. London, 1956.

Heyman, Jacques. "Gothic Construction in Ancient Greece." *Journal of the Society of Architectural Historians,* XXXI (March 1972), pp. 3–9.

Krautheimer, Richard. *Rome: Profile of a City 312–1308*. Princeton, 1980.

Landels, John G. *Engineering in the Ancient World*. Berkeley, 1978.

MacDonald, William L. *The Pantheon: Design, Meaning and Progeny*. Cambridge, MA, 1976.

Mango, Cyril. *Byzantine Architecture*. New York, 1976.

Mark, Robert. *Experiments in Gothic Structure*. Cambridge, MA, 1982.

Mark, Robert. *Light, Wind, and Structure.* Cambridge, MA/New York, 1990.

Mark, Robert and William W. Clark. "Gothic Structural Experimentation." *Scientific American,* 251 (November 1984) pp. 176–185.

Middleton, John Henry. *The Remains of Ancient Rome,* London, 1892.

Palladio, Andrea. *The Four Books of Architecture,* trans. Isaac Ware, London 1738; reprint, New York, 1965

Panofsky, Erwin. *Abbot Suger: On the Abbey Church of St. Denis and its Art Treasures,* 2d ed. Princeton, 1979.

Penelis, George, et al. "The Rotunda of Thessaloniki: Seismic Behavior of Roman and Byzantine Structures." In R. Mark and A. S. Cakmak, eds., *Hagia Sophia from the Age of Justinian to the Present.* New York, 1992, pp. 132–157.

Robertson, D. S. *Greek and Roman Architecture,* 2d ed. Cambridge, 1943.

Shelby, Lon R. *Gothic Design Techniques: The Fifteenth-Century Design Booklets of Mathes Roriczer ans Hanns Schmuttermayer.* Carbondale, 1977.

Shelby, Lon R. and R. Mark. "Late Gothic Structural Design in the 'Instructions' of Lorenz Lechler." *Architectura,* 9, 2, (1979), pp. 113–131.

Sear, Frank B. *Roman Architecture.* London, 1982.

Serlio, Sebastiano. *Tutte opere d'architettura e prospettiva di Sebastiano Serlio,* 1619; reprint, Venice, 1964.

Strayer, Joseph. *On the Medieval Origins of the Modern State.* Princeton, 1970.

Viollet-le-Duc, Eugène. *Dictionnaire Raisonné de L'Architecture Française du XIe au XVIe Siècle,* 10 vols. Paris, 1854–1868.

Ward-Perkins, J. B. *Roman Imperial Architecture.* New York, 1981.

Wren, Christopher. *Parentalia,* London, 1965; reprint, 1750.

4 Vaults and Domes

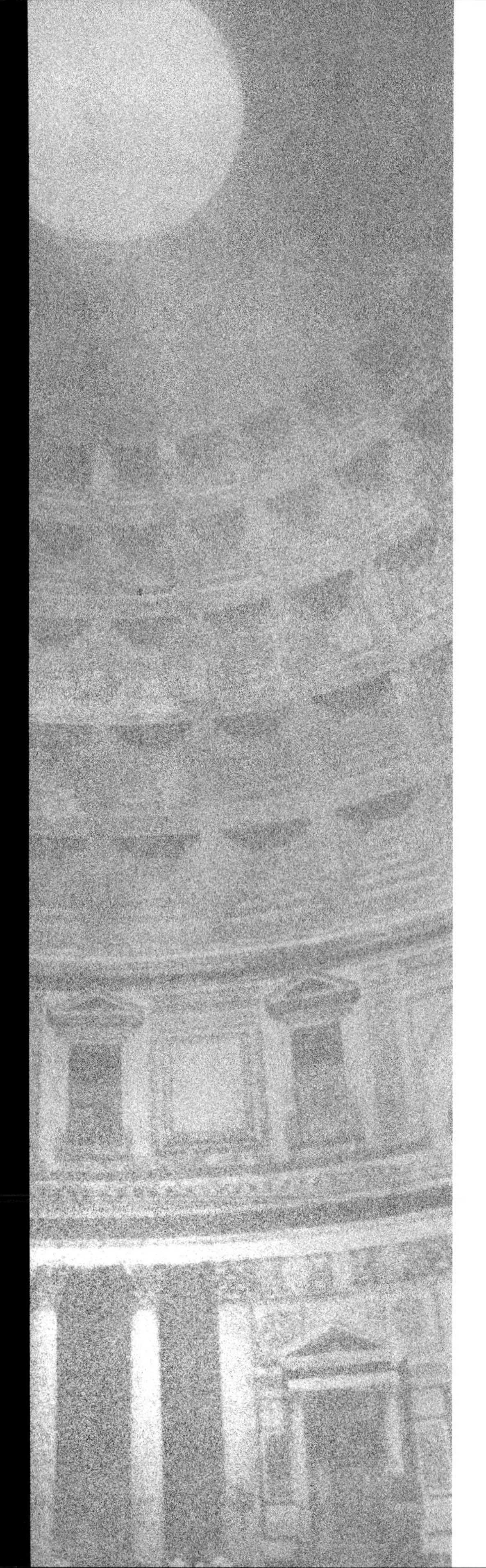

From early Mycenean *tholos* tombs to monumental imperial Roman structures, soaring Gothic cathedrals, and large-domed Renaissance basilicas, a masonry dome or vault has traditionally provided the architectural *summa* for buildings dedicated to important religious and political functions. The symbolic meaning of these forms, such as the physical representation of the "dome of heaven," the cosmos, or associations with palatine courts and kingly authority, figure prominently in many histories of architecture (see, for example, Smith, 79–94). Even where such associations have not been paramount, the desire to represent authority and/or belief has almost always been associated with sizable vaulted structures.

Early builders seem to have grasped the fundamental structural principles of masonry vaults and domes from their experience with simple, planar voussoir arches (chapter 3). All of the basic Roman vault and dome forms, for example, are related to the semicircular arch: to cover large spaces, Roman builders created barrel vaults by effectively aligning a sequence of parallel semicircular arches along a common longitudinal axis, groin vaults by intersecting two barrel vaults at right angles, and hemispherical domes by rotating semicircular arches about their centers (figure 4.1). But whereas a barrel vault loaded by deadweight behaves like a planar arch, the structural behavior of groined vaults and domes is far more complex, particularly if the material from which they are constructed behaves as a monolith (as does modern steel-reinforced concrete).

When built of unreinforced masonry, which is likely to exhibit cracking in regions subjected to even low levels of tensile stress (see chapter 1), vaults—and even more so, domes—tend to behave like an agglomeration of planar arches. The degree

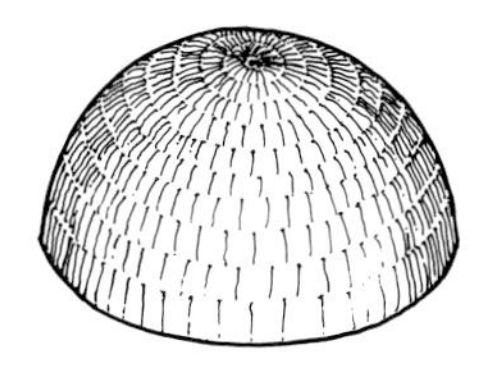

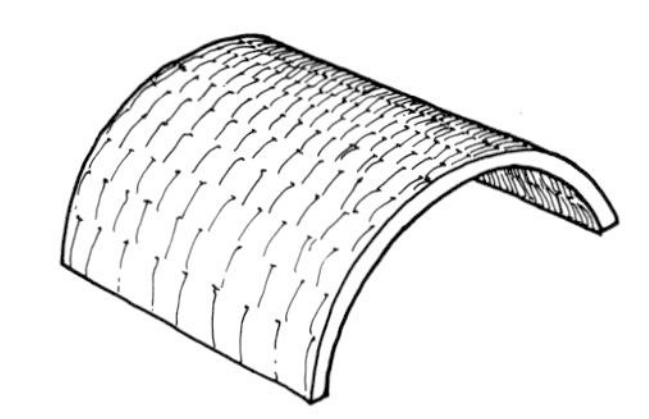

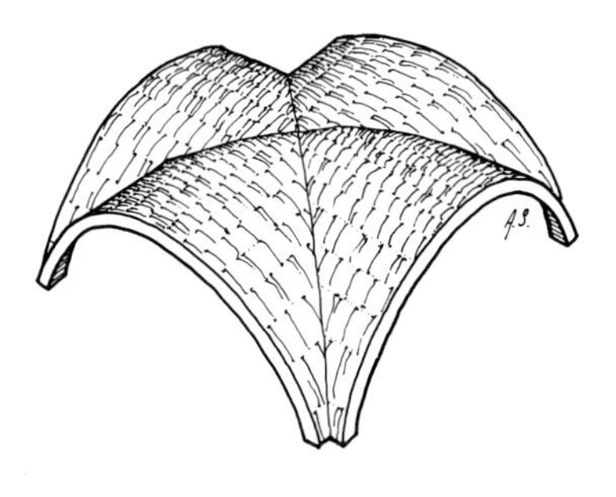

4.1 *Shell forms generated from a semicircular arch.*

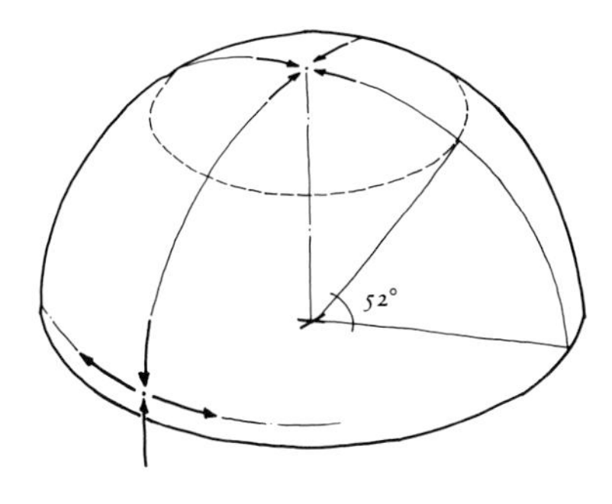

4.2 *Hoop tension and meridional compression in a monolithic, hemispherical dome.*

of such cracking can change drastically the distribution of forces exerted on their supporting structures. Take the case of the hemispherical dome under its own deadweight. A dome tends to deform downward at the crown and bulge outward at its haunches, in a manner similar to that of an arch. This spreading of the haunches stretches the fabric of the dome along its circumference, producing tension (known as "hoop tension"—similar tension would be experienced by a rubber band placed around the dome's expanding circumference). If the dome fabric can contain this tension without cracking (i.e., remain a monolith), little outward force will be exerted on the structure supporting the dome. In an unreinforced hemispherical masonry dome, however, hoop tension usually precipitates radial cracking, beginning at the dome springing and propagating upward toward the crown—stopping only when reaching a region of compression, about 52 degrees above the vertical, as shown in figure 4.2. After cracking, the dome behaves like an array of wedge-shaped arches (see comments on the Pantheon, below), and like an arch, it is entirely dependent for its stability upon the firmness of the abutments. The piers or walls on which it rests must resist the horizontal thrusts to prevent its spreading and collapse. For an unreinforced masonry dome, therefore, the concern of the designer was not with the dome alone, but with the vital interaction between the dome and its supporting structure.

As a matter of fact, all large-scale masonry vaults and domes share the common necessity of being provided with sufficient structural support even though their configurations have altered over the almost two-millennia span of time treated here. Several strategies were employed by the early builders to deal with the general problem of abutment.

Outward vault and dome thrusts, for example, can be countered with massive buttresses of sufficient weight to effectively nullify the effect of the thrusts (see figures 3.6 and 3.9). This approach is commonly found in Roman architecture, but as discussed in chapter 3, great masses of masonry may also overtax foundations and reduce the size of openings available for interior lighting, as well as escalate building costs. These pitfalls may be circumvented by providing *local* bracing in the form of wall buttresses, spur walls, or exterior flying buttresses. Another approach is to reduce thrusts by decreasing the weight of the vault or dome structure with lightweight materials and/or hollow building elements, a particularly effective strategy when applied to regions of the vault or dome furthest from the supports. Raising the peak of the vault or dome structure (i.e., giving it a pointed profile) also tends to reduce horizontal components of thrust in the manner of pointed arches discussed in chapter 3—a solution advantageous in construction because the steeper, lower portions of a vaulted structure may not require centering for their placement. Finally, wrought-iron tension hoops, or "chains," were applied in later buildings to contain outward thrusts at the haunches of domes.

Iron cramps, used in antiquity for linking individual wall blocks together, were applied also to strengthen vault and dome structures. In large Byzantine buildings, for example, cornices at the level of the dome springing were sometimes tied with metal cramps (Butler, 62ff), but such reinforcement could not have effectively resisted the huge thrusts generated by very large domes. Direct reinforcement, consisting of ties, both of timber and of iron, to secure the springing of vaults, began to be used only in the later Middle Ages. But such ties were considered temporary, to be taken down when the newly constructed masonry had settled and the mortar had cured. In areas of seismic activity, or where poor foundation conditions prevailed, however, they were sometimes left in place; for example, at SS. Giovanni e Paolo in Venice (1260–1305). This practice might have suggested the application of permanent iron reinforcement for later domes. But before the Scientific Revolution, its use could not guarantee stability of a dome because of the builders' inability to specify required amounts of tensile hoop reinforcement, or for that matter, even to acquire the necessary quantities of forgings with predictable levels of strength. As the following accounts of individual buildings make clear, iron reinforcement was entirely determined by trial and (at times, costly) error.

Before discussing the structure of historic vaults and domes themselves, the vital role of temporary timber **centering** for their erection needs to be emphasized, as it was for planar arches (timber centering with laggings used for the support of a ribbed vault under construction are illustrated in figure 3.19). Indeed, the requirements for initial precision and maintaining the rigidity of the formwork during assembly, as well as for its ease of removal afterward, are even more critical for spatial (three-dimensional) than for planar (two-dimensional) structures. But in all other respects, the description of arch centering given in chapter 3 applies equally well to the centering for vaults and domes, including the classification of its two basic types: (1) supported directly from the ground, and (2) springing from an elevated masonry pier or wall. Traces of the second type of centering are usually marked by putlog holes in walls that were used to support scaffolding for the workmen (Fitchen, 17) and by masonry projections from which the centering was sprung.

4.3 *"Treasury of Atreus," Mycenae, ca. 1325 B.C.: section.*

Ancient

As noted in chapter 3, the voussoir arch was not adopted by the Greeks for monumental architecture; nor were its spatial derivatives. Instead of true voussoir vaults and domes, space was normally enclosed by timber-framed roofing systems. The corbel arch, however, was developed spatially in a similar way to the later Roman formation of barrel vaults and hemispherical domes: a form analogous to a barrel vault is used in the galleries at Tiryns, and a dome-like room is found in the so-called Treasury of Atreus at Mycenae (both dating from the thirteenth century B.C.). The latter, 14½ meters in diameter, was formed by corbeling ringed courses of masonry in igloo fashion (figure 4.3). The structure was built initially in an excavation and later buried. This helped insure its stability because the pressure of the surrounding soil maintains its structure in hoop compression. The cross-sectional profile of the Treasury dome is rounded somewhat, approximating a parabola instead of a simple cone, and along with the careful jointing of the masonry, demonstrates a high level of skill on the part of the masons.

The use of the heavy corbeling technique for the construction of a monumental subterranean chamber seems appropriate, especially if the chamber forms part of a defensive position, since the frictional forces obtained from its massive deadweight could serve to discourage efforts by sappers attempting to gain entry. The extreme weight of such structures as well as their limitations of scale, however, may have led Hellenistic builders to experiment with vaults composed of trapezoidal voussoirs more closely resembling true arches (Coulton, 154).

Imperial Rome

The transition from stone or brick masonry to cast concrete for primary load-bearing structures at the end of the first century is perceived as having precipitated a "Roman architectural revolution" that allowed Roman architects to develop new, three-dimensional vaulted and domical forms for covering large-scale spaces within public buildings (see, for example, MacDonald 1982, 41–46; Ward-Perkins, 97–120). Nero's Domus Aurea, or Golden House, constructed after the Fire of Rome in A.D. 64 by the architects Severus and Celer, has been aptly described as "the first building in which we can clearly and unequivocally observe the impact of this new vision . . . giving these [revolutionary] forms monumental expression within the best known and most discussed building of its day" (Ward-Perkins, 101). A room of singular interest, particularly in the context of this chapter, is the octagonal hall. To cover the space, a truncated concrete dome measuring 13½ meters across the (interior) flat sides of its octagonal base, supported on eight radially braced piers, 5 meters high and rising an additional 4½ meters to terminate in a 6-meter-diameter oculus. The dome's form is unconventional for Roman construction: flatter than a hemisphere, apparently because of the need for lighting the adjacent chambers (shown in figure 4.4). But this inconsistency of form, together with the general novelty of the dome design, indicates that Roman architects during the middle of the first century were not encumbered by fixed canons of design—at least for relatively small-scale domes. The array of radial walls behind the piers that help resist the outward thrusts of the dome indicates as well a high level of structural sophistication.

The Pantheon, constructed on the site of an earlier temple in A.D. 118–128, is usually taken as the high point of the "Roman architectural revolution." Its 43-meter clear span (figure 3.25), far greater than that of any earlier known domed building, remained unmatched until the construction of the dome of Florence Cathedral some 1,300 years later. As discussed in chapter 3, a common mode of Roman wall construction consisted of placing a core of concrete between permanent brick forms. Domes, by contrast, were not usually faced with brick. The concrete dome of the Pantheon, constructed of gradated lightweight aggregates to reduce weight, and hence thrust, was likely formed over timber centering supported from the ground.

Stemming from the belief in the monolithic character of Roman pozzolan concrete, the structure

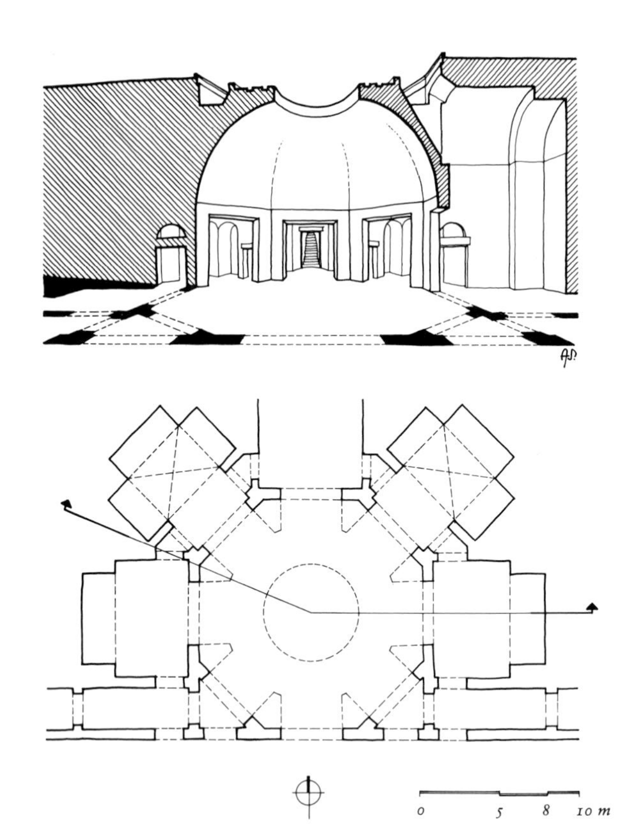

4.4 *Domus Aurea, Rome, A.D. 64–68. Octagonal atrium: analytical drawing and plan.*

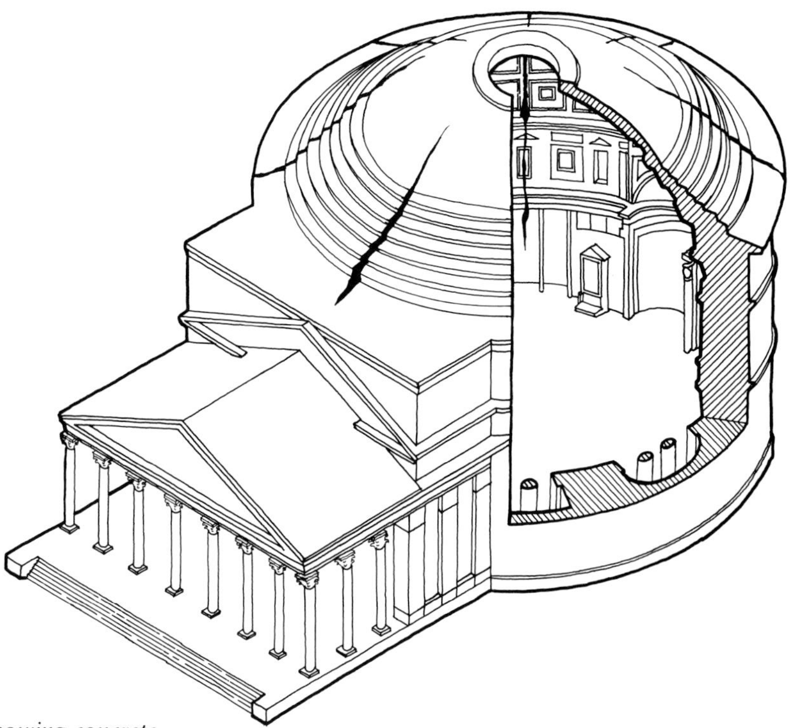

4.5 *Pantheon: "cutaway" view, showing concrete dome and wall and typical tensile crack patterns.*

of the Pantheon has been likened to that of a great metal pot lid (Middleton 1892, I:66 and II:131). Moreover, the concentric stepped rings near the base of the dome, which effectively thicken the shell, have been interpreted as giant reinforcement hoops (figure 4.5), allegedly functioning like metal chains in reducing tensile hoop stresses caused by the tendency of the base to splay outward under the tremendous dead weight of the dome. Yet, a numerical computer-model study (Mark and Hutchinson, 26), together with data from an earlier archaeological investigation (described in figure 4.6), revealed the extensive radial cracking in the dome and the supporting wall illustrated in figure 4.5. Because of the cracking, the stepped rings cannot possibly act as tensile hoop reinforcement, but their extra weight near the supporting walls is advantageous in helping to maintain the (cracked) arch segments of the dome in compression along meridians. The rings, in fact, represent the omnipresent stone surcharge of a typical Roman arch. While the Pantheon is a revolutionary building in terms of architectural form, it does not embody any revolutionary structural conception.

For all intents and purposes, then, unreinforced Roman concrete in large-scale domes could not guarantee appreciable resistance to tension. Moreover, the Roman builders seem to have understood this, as their measures taken to assuage problems caused by low tensile resistance attest: the use of lighter-weight aggregates (brick and tufa) near the dome crown to help reduce the magnitude of the dome reactions, massive supporting walls to act as rigid abutments, and the placing of the stepped rings over the dome in the same manner as surcharge over an arch. Roman concrete, then, offered no significant structural advantage over conventional masonry construction of brick or stone. Consequently, one may conclude that the decision to employ concrete in large-scale Roman architecture was based mainly on economic and constructional factors rather than structural considerations. Placement of concrete was, no doubt, more quickly executed and less costly than

4.6 Structural modeling of the Pantheon (finite-element modeling).

Numerical finite-element *modeling used for computer simulation of full-scale structural behavior involves many of the same considerations as the physical modeling described in chapter 3. The configuration of the structure, however, is now defined as a series of numerical coordinates, forming a mesh that becomes the geometric model for the computer. A series of equations relating to both loading conditions and the properties of the construction materials are used to calculate the displacement of all the mesh points in order to obtain a displacement pattern for the entire structure. This pattern then gives, through equations of elasticity, the same type of information about the overall distribution of internal forces in the actual structure as photoelastic model testing (further details on the approach may be found in Cook).*

For the Pantheon structure, a typical meridional section of the dome and wall was specified for modeling. Because of extensive openings for recesses and passages in the actual wall as well as coffering of the lower dome surface, no "typical" section truly exists. But we inferred that the illustrated model section, having a smooth interior surface and a 5.5-meter-thick solid cylindrical wall provides, for computational purposes, the same overall structural behavior as the actual dome on its 6-meter-thick voided wall. The varying density of the aggregates in the dome was, however, accounted for.

A first series of model tests was based on the assumption that the entire structure exhibits monolithic behavior (i.e., no cracking). Extensive regions subject to low tensile hoop stresses were found as well as the surprising result that these decreased in magnitude by about 20 percent with the step rings removed from the dome.

A second series of tests, more consistent with modern concrete design theory, assumed that the unreinforced concrete could withstand no tensile stress. The model simulation in this case is more complex than the first because the computer must now follow the effect of the propagating cracks in regions subject to tension. Extensive meridional cracking was indicated through the walls and extending up into the dome to an angle of 54 degrees above a horizontal plane passing through the dome center (Mark 1990, 65). Moreover, this result agreed almost exactly with crack length measurements taken while repairs were being made to the dome in the early 1930s (Terenzio, 280–285).

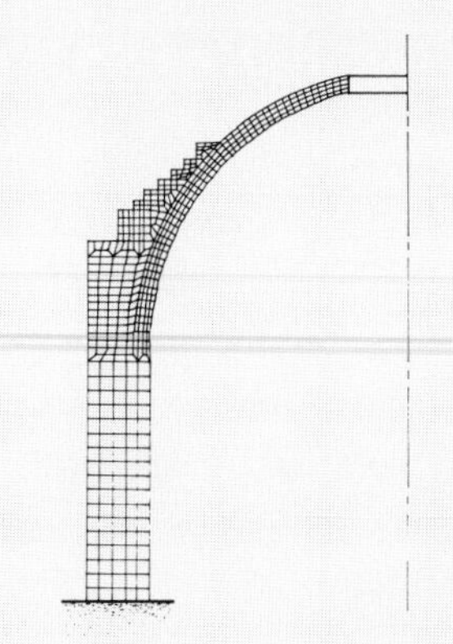

4.7 *Markets of Trajan, Rome, market hall: groined vaulting.*

that of brick or quarried stone. But rather than representing a "revolutionary" break with earlier structural tradition, the basic form and support of monumental concrete Roman buildings derived mainly from previous building experience.

A nearly contemporaneous example of Roman early concrete groined vaulting is found in the main market hall of the Markets of Trajan (figure 4.7). The major benefit of groined vaulting derives from its ability to "focus" the dead-weight forces within the vaulting to specific points of support. There are three components of the focused force resultants on an interior support: (1) a downward vertical component, from portions of the vaulting in adjacent bays, totaling half the weight of a full bay and transmitted directly to the piers below; (2) an outward-acting horizontal component that must be braced laterally, usually by massive piers or lateral walls; and (3) a longitudinal, horizontal component acting against the adjacent bay of vaulting, and usually stabilized by it. Exterior walls directly below the vaulting provide no support and can be entirely dispensed with, allowing large windows to be inserted. Thus, unlike a Pantheon-type dome that requires massive load-bearing walls around its perimeter, the vaults of the market allow the entry of light and the circulation of air into what must have been a crowded, busy space.

The hall is covered by six bays of rectangular-plan vaults with typical bay dimensions of about $5\frac{1}{2} \times 9$ meters (clear span) and 12-meter height from the vault crown to the floor below. Tiles, rather than timber laggings, were probably used on the centering of the market hall vaults to minimize carpentry and increase the speed of construction (MacDonald, 159).

Groined vaults achieved greatest expression in the baths and basilicas of the late Empire. The best-preserved example, from which it is still possible to capture the experience of Roman vaulted space, dates from two centuries after Trajan's Market: the triple vaults of the frigidarium of the baths of Dio-

4.8 (Opposite page) *S. Maria Degli Angeli (Baths of Diocletian), Rome: groined vaulting, ca. 300.*

4.9 (Above) *"Temple of Minerva Medica" (Pavilion in the Licinian Gardens), Rome, ca. 320: reconstruction (after MacDonald).*

4.10 (Right) *"Temple of Minerva Medica," recorded by Franz Innocenz Kobell in 1780.*

cletian in Rome, constructed originally between A.D. 298 and 306, and reused as a part of the present-day church of S. Maria Degli Angeli in a renovation by Michelangelo in 1561 (figure 4.8). All three of its rectangular-plan vaults rise to about 22½ meters and span 17 meters, but the vault over the central bay is almost a full meter wider than the 11-meter-wide end vaults. The substantial horizontal thrusts generated by the vaults are resisted within the building structure by heavy lateral walls (portions of these can be seen in figure 4.8), providing no obstruction to the large clerestory windows that so effectively light the interior.

The clear span of the contemporary basilica Nova (figures 3.29–3.32) exceeded 25 meters. Square-plan groin vaults rising 35 meters above the floor afforded large well-lit clerestories, while the flanking exedrae, having the same span as the high vaults, were covered with transverse barrel vaults 15 meters deep. Apart from fallen fragments (the structure seems to have collapsed in an earthquake in 1349), only the three exedrae from the north side of the basilica are still extant, but these serve to express the visual power of the original structure (Rivoira, 214). They also display the requisite surcharge that we have come to expect above Roman vaulting.

The fourth-century so-called temple of Minerva Medica in Rome (figures 4.9 and 4.10) marks the later Roman progression of Pantheon-related domed buildings to more skeletal construction. As discussed in chapter 3, this endeavor to incorporate large windows and skeletal support into a centrally planned building was not wholly successful, in that the structure supporting the dome had to be rein-

forced. Except for its not having a central oculus, the outside profile of its nearly semicircular 25-meter-diameter dome, whose crown rose 29 meters above the floor, resembled that of the Pantheon (including step rings). But the use of brick ribs in the dome (seen in figures 3.34 and 4.10) represents a later advance in Roman building. Rather than fully erecting the ribs as a skeletal first stage of dome construction, they were likely raised together with the light-tufa concrete infill; and at various stages of the construction process, the unfinished, truncated dome was capped by continuous rings of large tiles. Similar ribbing had been used in the vaults of the baths of Diocletian and the basilica Nova, but in none of these structures do the ribs seem to have been considered structural (as opposed to constructional) by their designers, since many end abruptly before reaching the crown or are otherwise interrupted (Ward-Perkins, 434–435). As will become evident, both aspects of the form and the construction details of this Roman pavilion are reflected in later imperial building in Constantinople.

Byzantine

The sixteen-sided brick dome of the church of SS. Sergius and Bacchus, though referred to as a "pumpkin dome" because of its cusped perimeter (figures 3.37 and 3.38), performs structurally in a similar way to a conventional hemispherical masonry dome. The base of the 16-meter-diameter dome, which rises to 22 meters above the floor, is pierced with eight small windows, offering a foretaste of effects pursued on a vastly larger scale in the greatest of all Byzantine churches.

The windows arrayed about the base of the 31-meter-diameter central dome of the Hagia Sophia in Constantinople, produce a ring of diffuse light and create the illusion that the dome is mysteriously suspended above the vast interior of the church (see figure 3.1). The dome, supported by four great arches as well as pendentives backed by massive surcharges (figure 4.11), is partially braced to the east and west by great semidomes (as illustrated in figure 3.40). The pendentives have the form of equilateral triangles on a common sphere whose equator adjoins the lower angles of the pendentive surfaces (figure 4.12). The diameter of this sphere, taken as the length of a diagonal across the inside corners of an opposite set of piers, is 46 meters.

Although archeological evidence for the original dome that collapsed in 558 is unavailable, sixth-century descriptions indicate that the dome interior was likely profiled from the same spherical surface as the pendentives (forming a so-called pendentive dome). Such a dome would have had an interior radius of 23 meters, making it 6 meters lower than the present dome, which agrees with the height difference between the first and second dome cited by ancient chroniclers (Procopius). No further direct evidence for a reconstruction of the first dome exists, but a comparison of the cross-sections of the Hagia Sophia (taken across the diagonal illustrated in figure 4.12) and the Pantheon suggests that Justinian's builders were referring consciously to the Pantheon in their design, lending further credence to the proposed pendentive dome profile. The lower portions

4.11 *Hagia Sophia: central dome, base, and western semidome, from southwest minaret.*

4.12 Transitions: Squinches and pendentives.

Squinches, corbeled or arcuate, fill in the upper corners of a square-plan or polygonal-plan space to provide transition to a polygonal-plan or round-plan base. Pendentives *are triangular segments of vaulting at the base of a dome, usually for transition from a round-plan to a square-plane base. The stability of a pendentive of masonry, in a similar way to that of a planar, semicircular arch, is dependant on the surcharge behind the pendentive surface.*

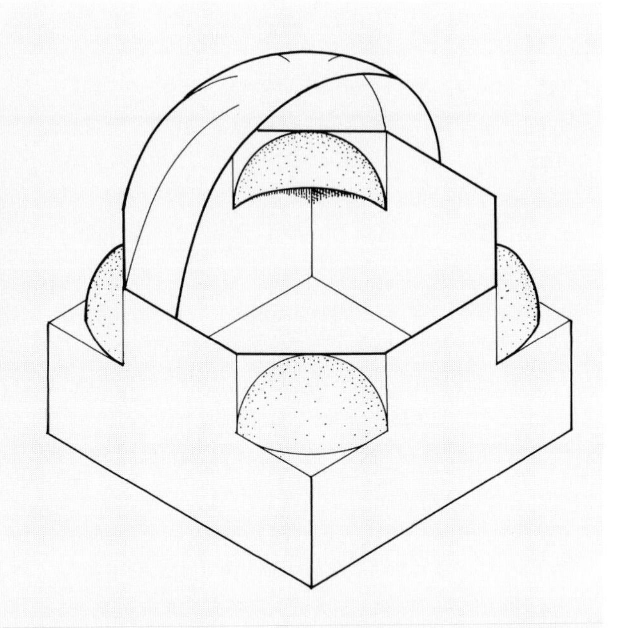

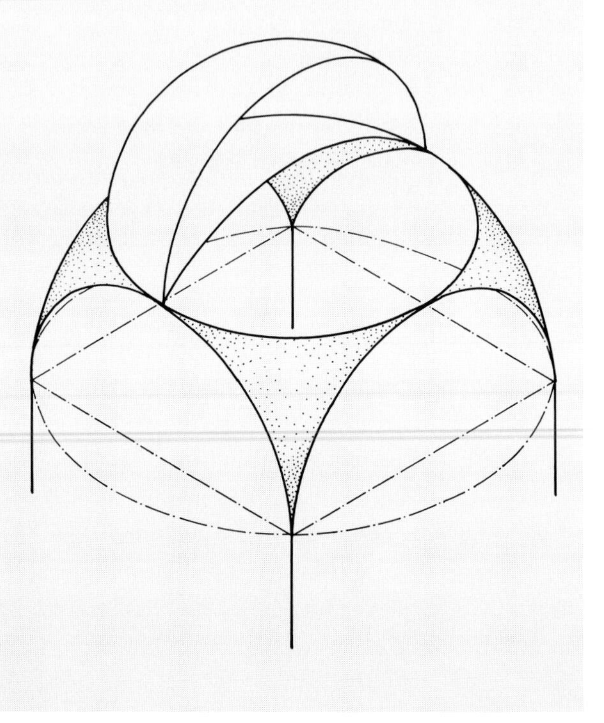

of both domes (shown in figure 4.13) are similarly massive and provide support to lighter "shells" above, although in the Hagia Sophia this massiveness is concentrated in regions above the four piers. Both "shells" also subtend angles of about 90 degrees and are of almost the same span. Hence it may be inferred that the scale of the first dome of the Hagia Sophia—as with the later, great Renaissance domes—was suggested by the model of the Pantheon.

It has been traditionally believed that the designers, Anthemius and Isidorus, incorporated the windows at the base of the dome solely for visual effect. Indeed, the placement of windows in a region where tensile hoop forces would be expected to be critical in a monolithic dome seemed audacious, even unwise. However, the Pantheon has shown that even a massive dome made of cast material does not behave in monolithic manner—meridional cracks cause it to act as a series of separate wedges. The builders of the Hagia Sophia must have understood that their dome, thinner than that of the Pantheon and supported on arches and pendentives rather than a practically continuous vertical wall, would be prone to the same type of meridional cracking as that experienced by the Pantheon. The window openings were thus both a startling visual element and a prudent expedient to ward off cracking. Although the actual form of the original openings remains unknown, one may hypothesize that it was similar to the present window configuration with the openings carried to an angle exceeding 50 degrees above the horizontal equatorial plane. This level would have been roughly equal to the crack length experienced by the Pantheon and suggests that the designers had knowledge of earlier dome fractures.

As remarkable as the dome with its ring of windows is the supporting structure of arches, semidomes, and piers. A clear understanding of the structural action of this supporting fabric proves difficult to obtain, not only because of the complex interaction between dome, semidomes, arches, and piers, but also because of the large deformations within the building, in many instances discernible to the naked eye (see, for example, figure 3.36). The most significant evidence of the great forces at work in the structure are the inclinations of the four main piers that tilt outward along the directions of both axes. This deformation is caused by a combination of outward thrusts from the dome, but even more so from the four great arches (and their surcharges) that connect the piers and support the base of the dome (Mark 1990, 88). The outward tilt of the piers is greater in the lower regions, indicating that the upper portions were probably erected in plumb after some initial deformation had occurred below. Their pres-

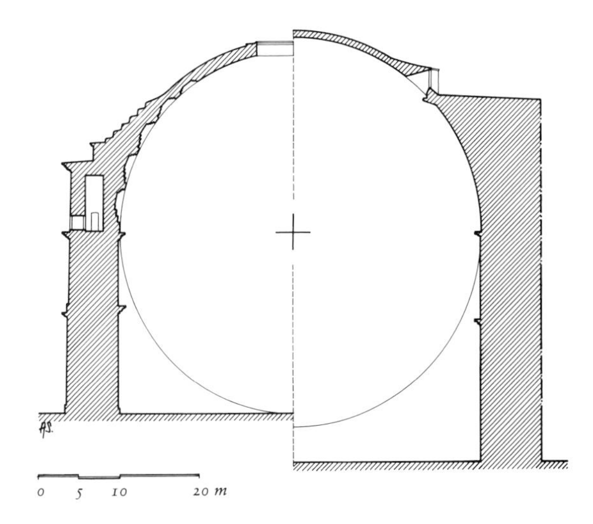

4.13 *Comparative diagramatic cross-sections: Hagia Sophia with first dome (partial section along diagonal) and (left) Pantheon (after Mark).*

ent inclination in the north-south direction exceeds that in the east-west direction by a factor of about two below gallery level, and about three at the gallery level and above (Mainstone 1988, 87). Deformation also manifests itself in the lack of symmetry in the arches spanning across the aisle and gallery walkways. The dome base, originally intended to be 100 Byzantine feet on each side, is no longer square.

Although the dome of the Hagia Sophia remains the crowning structural achievement of the Byzantine east, technological development occurred also in the western part of the empire: namely, the introduction on the Italian peninsula during the fifth and sixth centuries of hollow terra cotta tubes. As already noted, Roman builders had recognized that the deadweight of a dome or vault has a direct effect on both horizontal thrust and overall stability. This knowledge probably led later Italian builders to adopt hollow building materials to lighten a variety of vaulted structures. Although masonry construction is weakened by using hollow elements, the reduction in deadweight lessens forces throughout the entire structural system. Because the primary problem in vault and dome design is to contain the outward thrusts, lessening the dome weight allows lighter supporting walls and places less weight on timber centering during construction.

An early example of a hollow-tube dome is found in the Orthodox Baptistery in Ravenna, the western capital of the late Roman Empire. Installed around 450, its $10\frac{1}{2}$-meter-diameter dome, rising $14\frac{1}{2}$ meters, contains terra cotta tubes similar to those used in water conduits, laid in multiple rows about the circumference. Laying the tubes horizontally in this manner avoids the constructional difficulties of a radial layout that would create the need for odd wedge-shaped sections as the circumference narrows toward the crown. Also, the builders wanted to provide structural continuity about the perimeter of the dome, most likely to avoid cracking of the interior surface intended for decorative mosaics.

The Justinianic Church of San Vitale in Ravenna (figure 3.41), a contemporary of the church of SS. Sergius and Bacchus (and of Hagia Sophia), shares many of that building's peculiarities, including a domed central core structure supported on eight columns as well as the technique of employing long, thin bricks that characterized building in the Byzantine capital. The vaults and $16\frac{1}{2}$-meter-diameter central dome that reaches a height of 28 meters, how-

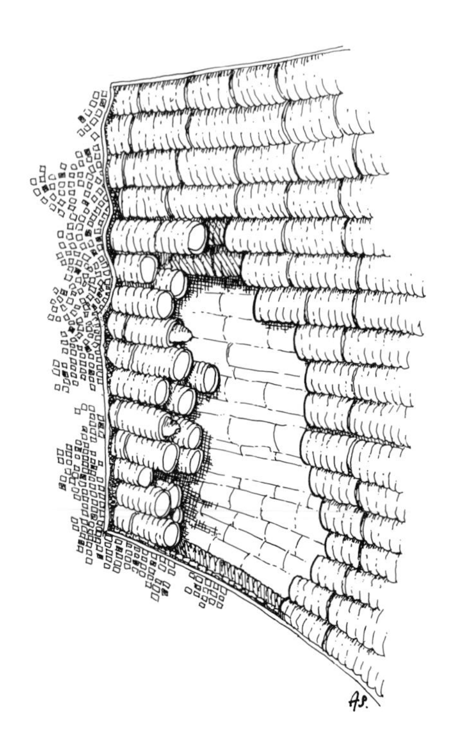

4.14 *Hollow vessels used in dome construction.*

ever, contain hollow tubes, a purely western technique (figure 4.14). Freed from the large buttresses that would have been required to counter the outward thrusts from traditional heavier vaulting, the builders of San Vitale were able to employ large clerestory windows, allowing more light to enter the building and enhance the interior surfaces, a most important factor given the huge sums spent on decorative mosaics.

It should be noted that the San Vitale chapel dome is not a true dome; rather it is an octagonal cloister vault. Yet it is sufficiently thick so that the horizontal rings of a true masonry dome are imbedded within its octagonal internal and external profiles, allowing the structure to behave much like a true dome (as discussed in greater detail with regard to the construction of the dome of Florence Cathedral, below).

Early Medieval

Despite the clear indebtedness of Charlemagne's Palatine Chapel at Aachen to San Vitale, including the similarity of scale (its dome is but two meters less in diameter and two meters greater in height than the dome of San Vitale), the entire chapel, including the dome, is constructed from ashlar rather than terra cotta (figure 4.15). The dome is supported by a smaller drum than at San Vitale, and restrained against outward spreading by a radial array of inclined barrel vaults. As noted in chapter 3, the heavier construction of the chapel, stemming from the need to buttress its more massively built dome, served also to change the architectural character of the building. These aesthetic qualities were to remain an important factor in building in general, and vaulting in particular, throughout most of western Europe, where builders continued the practice of monumental building in stone over the following centuries.

Romanesque

As at Aachen, the influence of earlier Byzantine architecture can be seen in the late-eleventh-century design of the basilica of St. Mark's in Venice. The interior space of St. Mark's is covered by five brick-domed bays in a Greek-cross plan, resting atop pierced drums that bring light to the upper reaches of the church. The central dome, rising 29½ meters above the floor and just short of 13 meters in diameter, is some 2 meters greater in both height and diameter than the four subsidiary domes. Poor construction of the supporting piers (see chapter 3) may have necessitated an innovative structural solution when the exterior silhouette of these domes was modified in the thirteenth century (figure 4.16). Their charming profile, still seen today from the piazza, is defined by domes and lanterns constructed not of stone but of lead-sheathed timber. These relatively lightweight structures (figure 4.17), rising 10 meters above the interior brick domes, helped relieve the pier footings placed on sandy Venetian subsoil, which might have been overburdened had the outer domes been constructed of brick. For similar reasons, lead-sheathed timber double domes were used in a number of later Venetian buildings. And, as will be developed later in this chapter, the Venetian double-dome scheme may have led to the strategies of large dome construction employed well into the modern era.

4.15 *Palatine Chapel, Aachen: cloister vault.*

4.16 *St. Mark's, Venice: thirteenth-century timber roof domes.*

4.17 *St. Mark's, Venice: cross-section through the domes (after Cicognara, et al.).*

Even after the advent of the crusades in 1095, Romanesque architecture continued to remain regional. Yet the domed-bay construction of St. Mark's found a following in a number of churches of the Aquitaine that reached as far north as the Loire Valley (at Fontevrault Abbey). The domes over the nave of the church at Souilliac (ca. 1130), for example, supported by arches and pendentives that carry the dome forces to stout piers, allow for clerestory windows in the tympana below (figure 4.18). Perhaps because such domes divide the nave into separate units—instead of creating a unified space leading to the altar—large vaulted Romanesque churches in the West (in contrast to those possessing the open timber roofs discussed in chapter 5) most often employed barrel vaults with transverse arches, or ribs, for their high central vessels. But such vaults had an important structural limitation.

Recall that the simple barrel vault is derived from the linear translation of an arch; in effect a masonry barrel vault acts as an array of side-by-side parallel arches, and the presence of ribs does not alter this characteristic behavior. The ribs, no doubt, helped in the vault erection process, and they aid in visually defining the bays, but they perform little structural function. Hence a barrel vault, like an arch, normally requires continuous, rigid walls for support; and by the same reasoning, it cannot accept large window openings within the fabric of the barrel itself.

The church of St. Foy at Conques, whose nave of 21-meter height and slightly less than 7-meter span was vaulted between 1087 and 1095 (figure 3.48), provides a fine example of how Romanesque barrel vaulting was usually deployed as well as the attendant problem of lighting the upper reaches of a large church. Vertical loading of the vault is carried by longitudinal arches to the piers, and outward thrusts are resisted by structure in the galleries, as described in chapter 3. Daylight enters the upper space through window openings at the building's perimeter and passes across the galleries, indirectly lighting the central vessel. As a result, the vault intrados actually receives far less light than the nave below.

One of the important issues prodding the later development of Romanesque church architecture and of early Gothic architecture that followed was the quest to increase illumination levels high in the building. As described in chapter 3, the nave walls at the third abbey church of Cluny supported a pointed barrel vault of 10½-meter span that reached almost 30 meters above the floor—the highest of any Romanesque interior—and some 10 meters above the buttressing of the wall afforded by the side aisle vaulting (see figure 3.51). The resulting tall clerestory enabled the builders to open up the walls to admit direct light, but the combination of thrusts from heavy vaulting and unsupported upper walls proved perilous.

A far more stable, if novel, arrangement of barrel vaulting is employed in the nave of St. Philibert at Tournus. Instead of the customary alignment of the axis of the barrel vault with the main longitudinal axis of the church, each nave bay is covered with a *transverse* barrel vault, about 6½ meters wide by 5½ meters in length, as illustrated in figure 4.19. Rather than obtaining support from the nave walls,

4.18 *Souilliac (Aquitaine), ca. 1130: domed vaulting of the nave.*

these relatively light vaults (because of their shorter spans), which reach a maximum height of 23 meters above the church floor, rest on a series of deep transverse "diaphragm" arches that connect the main piers. Except at the ends of the nave, where more massive structure is available, horizontal thrusts from individual vaults are balanced by the opposing thrusts of adjacent vaults, as with the Roman aqueduct arcades discussed in chapter 3. The diaphragm arches then need carry only the vertically acting weight of the vaults. While these in turn thrust outward on the walls, their abutment is brought low enough on the walls to be resisted by structure under the side-aisle roof. The physics of the St. Philibert solution is reflected in its tall, relatively slender columnar piers that need carry only vertical loads, and in the fact that the tympanum walls at the ends of the vaults were safely pierced with windows to furnish direct light.

Although the transverse barrel vaults in the nave of Tournus appear to be a practical solution to the problem of supporting barrel vaults, while allowing relatively large clerestory windows, they had but slight effect on the building practice of the time. A number of factors have been suggested to account for this, including the possibility of the transverse arches having poor acoustical qualities, and the fact that they "disrupt" the visual continuity of the church interior in reference to the altar, similar to the domed bays discussed above. A more likely answer is simple chronology. These vaults are usually dated late in the building's construction, in the first quarter of the twelfth century, at the same time that ribbed quadripartite vaults had already begun to be employed for the main vessels of churches. By its ability to "focus" deadweight forces, in a manner

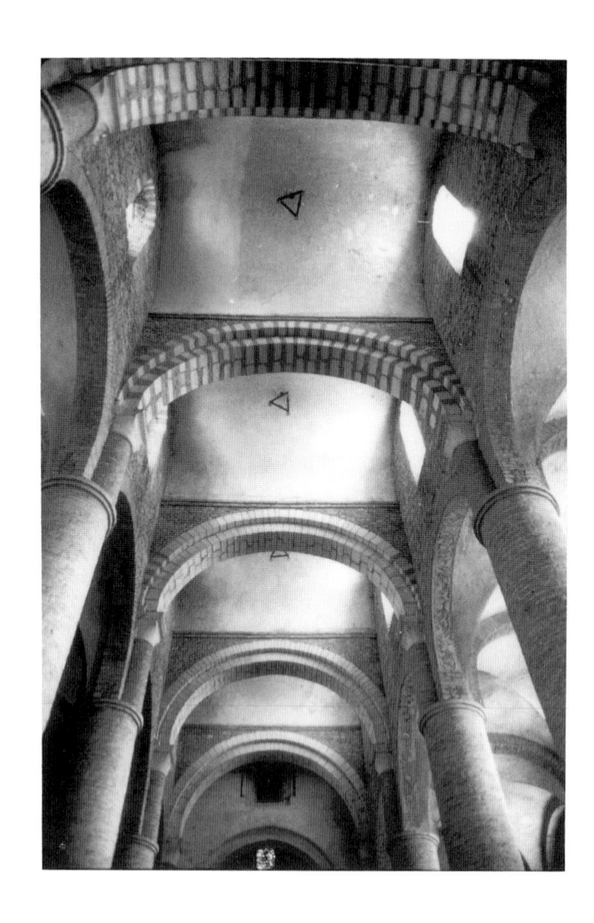

4.19 *St. Philibert, Tournus, ca. 950–1120: lateral, barrel vaulting of the nave.*

similar to Roman groined vaulting, ribbed vaulting too would free clerestory walls (the exterior walls below the vaulting) for fenestration.

Ribbed quadripartite vaulting appeared simultaneously in Italy and England around the turn of the eleventh century. England's Durham Cathedral, begun in 1093, was the first fully ribbed-vaulted building (figure 4.20). Although its oldest high vaults were replaced in the thirteenth century, the original ribbed vaults of the aisles, dated before 1100, remain intact; and the main vaults of the nave and transepts, completed by 1133, also retain their original form (Bony 1983, 7, 10). As already noted, ribbed vaults have all the structural advantages of groin vaults—supporting forces are focused on the piers at the corners of the bay, allowing the creation of an open clerestory below the vault—but they are also easier to construct (see chapter 3), which became especially important as nave vault heights increased. Moreover, the geometric pattern created by the ribs has inherent aesthetic appeal which, given the extensive geometric patterning of the nave piers, makes it likely that aesthetics also contributed to the adoption of ribbed vaulting at Durham.

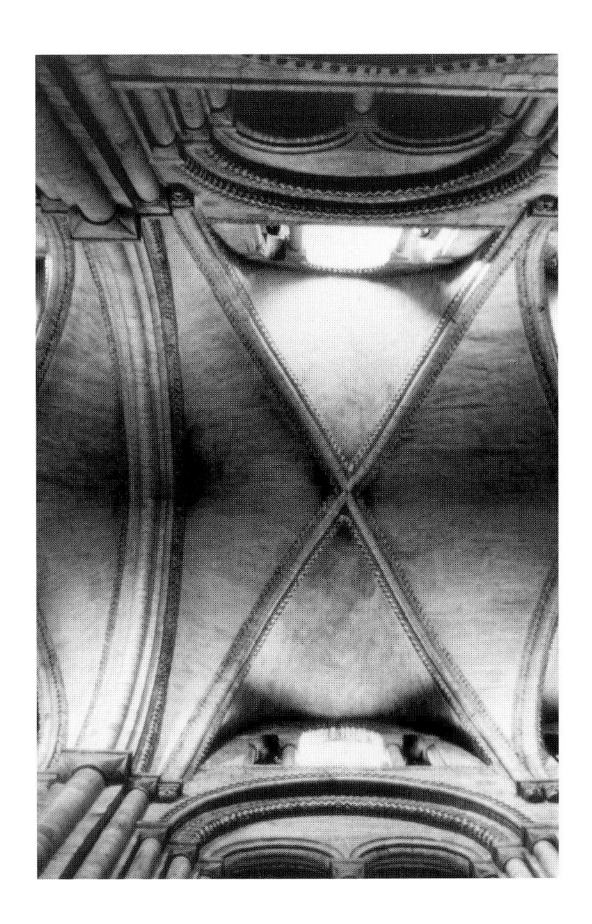

4.20 *Durham Cathedral, nave, ca. 1115–1133: early ribbed vaulting.*

Gothic

Pointed, ribbed vaulting entered the vocabulary of builders in the Ile-de-France region during the 1120s and 1130s, and it was soon thinned and lightened to allow the reduction of supporting walls (Bony 1983, 26–36). But the flying buttress was not introduced until almost a half century later, when lofty Gothic construction began to exceed the limits permitted by wall buttresses alone.

Abbot Suger's construction at St. Denis in the 1140s was at a relatively modest scale. The height of the ambulatory vaults does not quite reach 8 meters, and the maximum span between the piers is less than 3 meters. Yet the choir takes full advantage of the flexibility of the pointed arch to maintain constant vault height even in the irregular bays that curve around the apse. An array of "broken-rib" (free-rib patterned) pointed vaults serves to open up the walls, lighten supporting piers (by having vaults "lean against one another" to cancel horizontal forces upon the piers), liberate circulation, and create an ethereal ambience (figure 3.56). In some ways, the construction is even more remarkable for the fact that we have Suger's own description of the abbey (Panofsky).

By the mid-twelfth century, the almost ubiquitous groined vaults found over the side aisles of Romanesque churches were superseded by quadripartite ribbed vaulting. For the high central vaults of the larger churches, however, the early rectangular-planned quadripartite vaulting (as at Durham) gave way to square-planned sexpartite ribbed vaulting, each array covering two bays. With few exceptions, sexpartite vaulting was specified for the naves of the major Gothic churches of the second half of the twelfth century, including the cathedrals of Sens, Laon, Paris, and Bourges (figure 4.21). Then suddenly, after the start of the thirteenth century, all large Gothic churches, including Chartres Cathedral, illustrated in figure 4.22, deployed ribbed quadripartite high vaults.

Two essential questions have been raised about the technology of mature Gothic ribbed vaulting. The first concerns **the structural function of the ribs.** Is the weight of the vault supported by the ribs and transmitted by them to the piers below? Is it this function that allowed Gothic vaults to be so thin when compared with the heavy, unribbed groined vaults found in Roman buildings? The second query has to do with identifying the cause of the sudden transition from sexpartite to quadripartite vaulting around the year 1200.

Because of the striking linear articulation of vaulting surfaces, it had been generally assumed that Gothic ribs were intended to provide the primary support to the vaulting. Some observers, however, began to question this view when Gothic vaulting was shown to be stable even after its "supporting" ribs were blown away under bombardment during the First World War. The debate over the role of the ribs was polarized after the 1934 publication of a study in which it was asserted that forces within a vault (figure 4.23) follow the same path a ball would take when rolling along its surface (Abraham, 34). From this, it follows that the vault forces are attracted to the groins (folds in the vault above the ribs). But, it was argued, because of the local thickening of the vaults at the groins (shown in figure 4.23), the ribs themselves do not transmit the weight of the vaults.

More recent structural studies with both three-dimensional photoelastic (physical) models and finite-element (computer) models (of the type described in figure 4.6) of typical Gothic vaulting (figure 4.24) reveal that forces within the vault are not drawn to the groins, as postulated by Abraham, but rather follow the shortest paths directly toward the supporting piers, and the structural role of the ribs is actually slight. Essentially, the ribs support only

4.21 *Bourges Cathedral: sexpartite vaulting of the nave.*

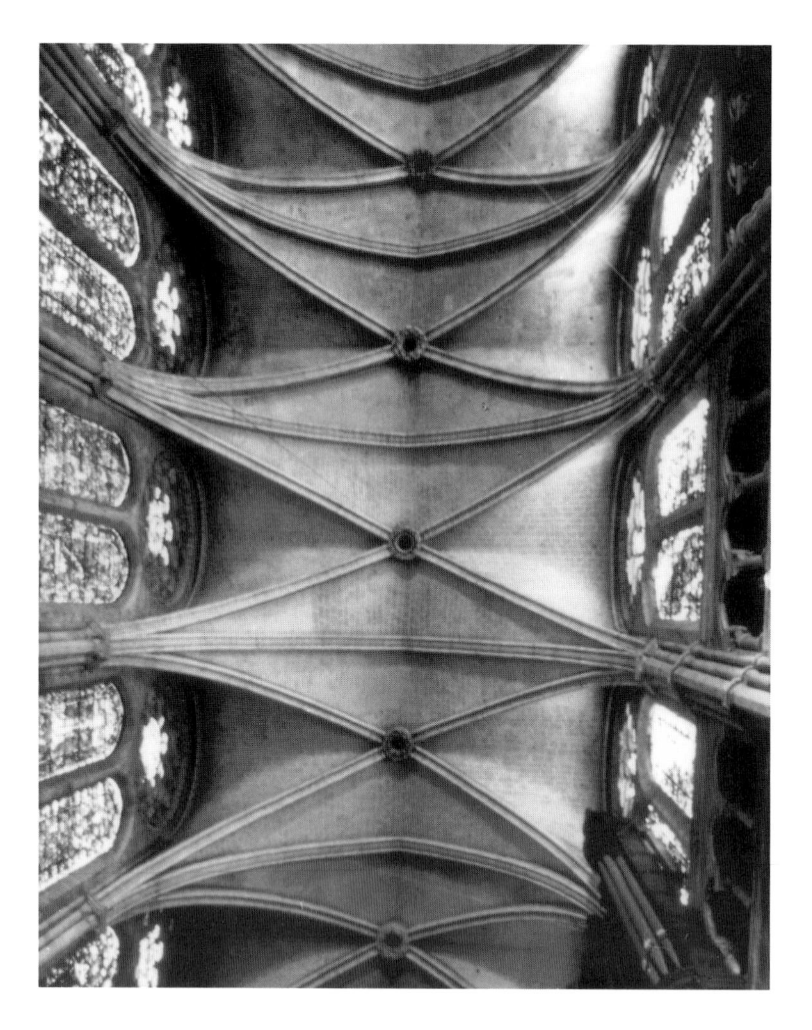

4.22 *Chartres Cathedral: quadripartite vaulting of the nave.*

4.23 *Groined-vault geometry and force distribution (Abraham).*

tation of radially ribbed vaulting to a square- or rectangular-planned bay involved some peril. The central opening created by the perimeter of the conoid tops requires closure by a large, flat surface which, because of bending, can be an incipient "Achilles heel" in this vaulting system. Bear in mind that despite possible cost gains from the use of standardized (radial) ribs, an account from the King's College Chapel quotes the cost of building the fan vault as being 67 percent greater than that of a conventional vault of the same size (Leedy, 29). The evidence indicates that it is not only in appearance that fan vaulting possesses luxuriance.

The dome of the Cathedral of Florence, Santa Maria del Fiore (figure 4.31), a structure poised between the Gothic and the Renaissance, was heavily dependant on the building tradition of Gothic master masons. Yet its architect, Filippo Brunelleschi (1377–1446), devised his solution for building the dome through the study of ancient Roman buildings.

Begun in 1294 by Arnolfo di Cambio, the erection of the cathedral was guided by a succession of master masons, with administrative continuity provided by the Wool Guild. In 1367, it was decided to enlarge the ongoing building project and to include a 42-meter-diameter cupola with a pointed profile—specifically, a *quinto acuto,* the form of a circular segment whose radius is one fifth greater than the span of the vault (see figure 3.63). One feature of the final dome design not specified in the 1367 building program was its double shell (figure 4.32), which affords access to the dome for maintenance by stairs located in the interstitial space without resorting to unsightly exterior scaffolding or stairs, a smooth continuous surface on both the inner and outer surfaces of the dome, and appreciable

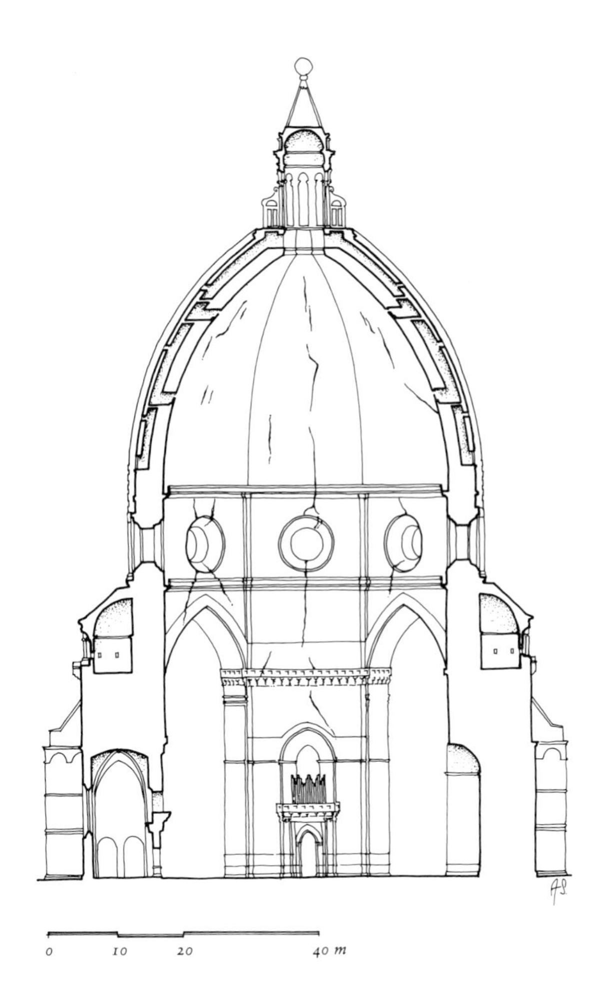

4.32 *Florence Cathedral: cross-section through the dome, showing typical tension crack pattern (after Superintendent of Architecture, Florence).*

reduction in deadweight from the large cavity created within the dome. Brunelleschi's cupola, however, is not light. The inner shell is over 2 meters thick. The lighter outer shell, intended mainly as protection against the elements and to provide the desired external architectural form, is about a third of the inner-shell thickness.

As in the case of the dome of the chapel at Aachen, the Florence dome is actually an octagonal cloister vault rather than a true dome. But by containing continuous horizontal rings of masonry within the perimeters of its flat "sails," the structure behaves much like a true dome; that is, the rings remain circular under compression and do not deform as would a thin octagonal shell. The angles at each of the corners of the octagonal plan are occupied by massive ribs, over 4 meters deep, extending through the entire dome and articulated on the exterior by a travertine profile. Their extreme depth also helps to contain the "line of thrust" from the dome, although probably just barely.

The span across the flat sides of the dome base almost exactly matches that of the Roman Pantheon. It was the first western dome to rival the span of the ancient temple, yet to Leon Battista Alberti, an even more incredible feat was Brunelleschi's construction of the dome without centering. Brunelleschi realized that by maintaining the horizontal rings of masonry within the perimeter of the truncated dome during construction (as discussed above), the structure would be self-supporting. Centering would not be required for the lower portion of the dome as long as construction proceeded in horizontal layers, so that an entire ring of masonry was completed before beginning the next one. As construction neared the crown and the working surface began to tilt inward toward the center, he used a *spinapesce,* or herringbone brick pattern—rectangular, flat bricks (similar to Roman bricks) laid with a "soldier," or upright brick, about a meter apart (figure 4.33)—to help hold the bricks in place while the mortar cured and the next segment of each new ring was being placed (Mainstone, 1977).

Reinforcing "chains," including six made up of sandstone blocks and four more of wood, were originally specified to be placed around the circumference of the dome (Saalman, 55–56). The stone chains were to be linked by iron cramps leaded into the stone in a fashion similar to Byzantine practice. The builders, however, must have realized that these would not be very effective, because the cramps holding the stone chains would likely bend and pull out under tension, and the capacity of wood relative to the great forces within the dome would be limited; only one of the wooden chains was put in place. Presumably Brunelleschi and the masters directing construction depended on the pointed Gothic profile of the dome and the massive ribs and drum to provide stability. Even so, cracks were recorded in the dome in 1639. Despite a long record of similar sightings, though, to date no major reinforcement has been added. Recent, continuous monitoring of the crack pattern shows major fissures running vertically from the top of the ground level arcade through the supporting structure and drum up to the compression zone near the crown of the vault (illustrated in figure 4.32). These cracks expand and contract seasonally, opening and closing in direct correspondence to the temperature of the masonry (Rapporto). Disturbingly, the seasonal opening and closing never quite returns the crack to its original position; rather, the crack grows incrementally each year. A critical question raised by these data—and which may be answered only after further observations and struc-

4.33 *Florence Cathedral:* spinapesce, *brick-herringbone construction (after Rossi). Tapering between projecting "soldier" bricks provides grip for the last brick layer until a new, complete ring has been laid.*

tural studies—is whether the crack growth is a relatively recent phenomenon, implying an increasing rate of expansion, or has been characteristic of the dome throughout its history, implying a slow, controlled expansion whose rate has remained constant.

Similar problems of masonry cracking and its consequences in terms of building endurance were to engage designers of the great Renaissance domes.

Renaissance

Interest in monumental domed buildings was revived by the architects of the Renaissance, but most of those projects were of small scale. Even so, when Pope Julius II decided to replace the Constantinian basilica of St. Peter's in Rome, Bramante produced plans for a domed building rivalling even the largest imperial Roman structures. Serlio's drawing of the Bramante design shows a single-shell dome, with step rings similar to those of the Roman Pantheon (figure 4.34). The structural viability of this design, however, is questionable since the step rings over the haunches of the dome are not as large as those of the Pantheon—and without enough deadweight in the rings to direct the thrust line down into the supporting piers, structural distress would likely have resulted. Yet more massive surcharge would probably have endangered the foundations, for the southernmost arches of the crossing were cracked before the dome was completed, probably because of differential settlement of the pier foundations.

For thirty years following Bramante's death in 1516, a succession of architects were appointed to direct construction, resulting in significant alteration of Bramante's original scheme. But with the

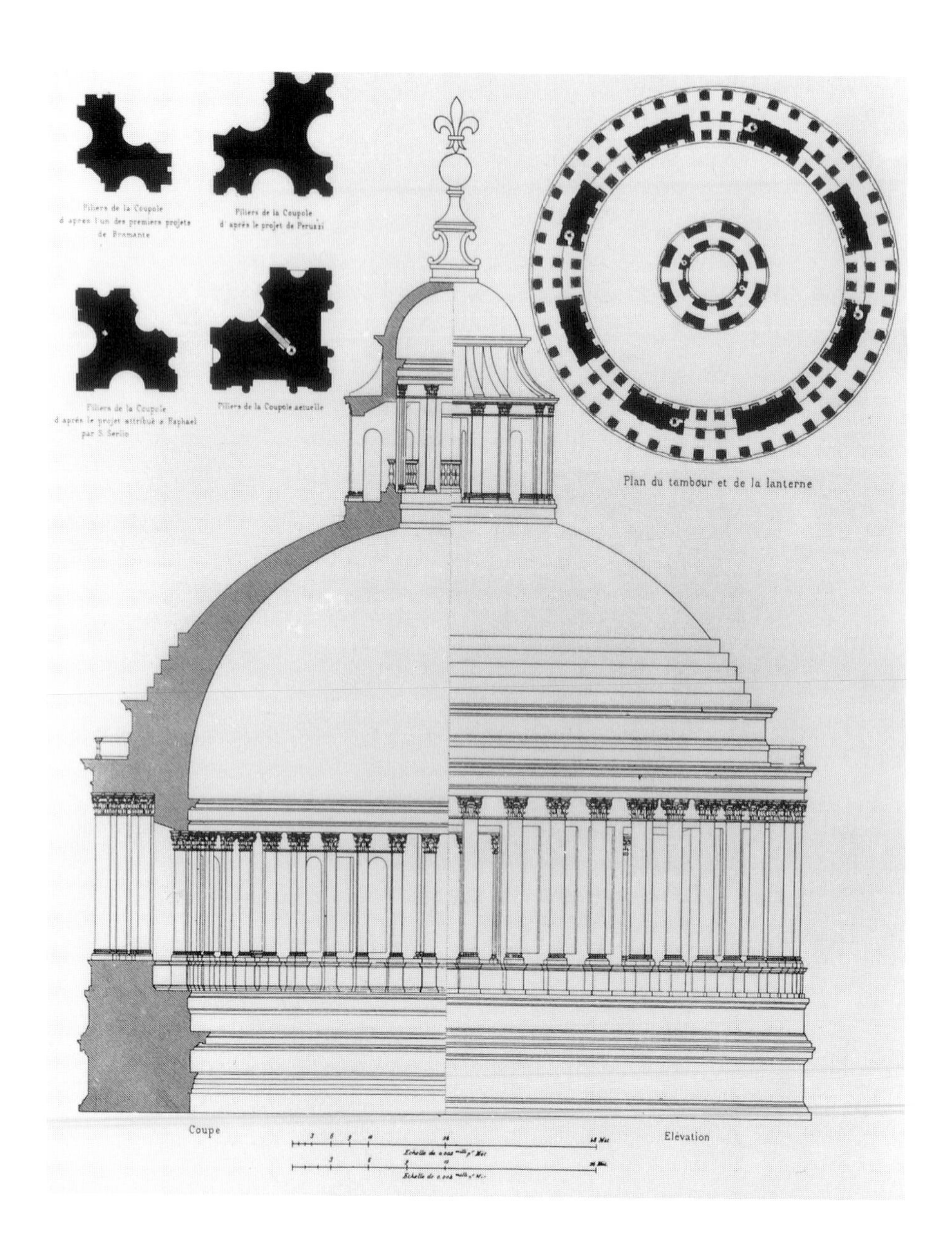

4.34 St. Peter's Basilica, Rome. Bramante design, ca. 1506, as illustrated by Serlio. Cross-section through the great dome (Letarouilly).

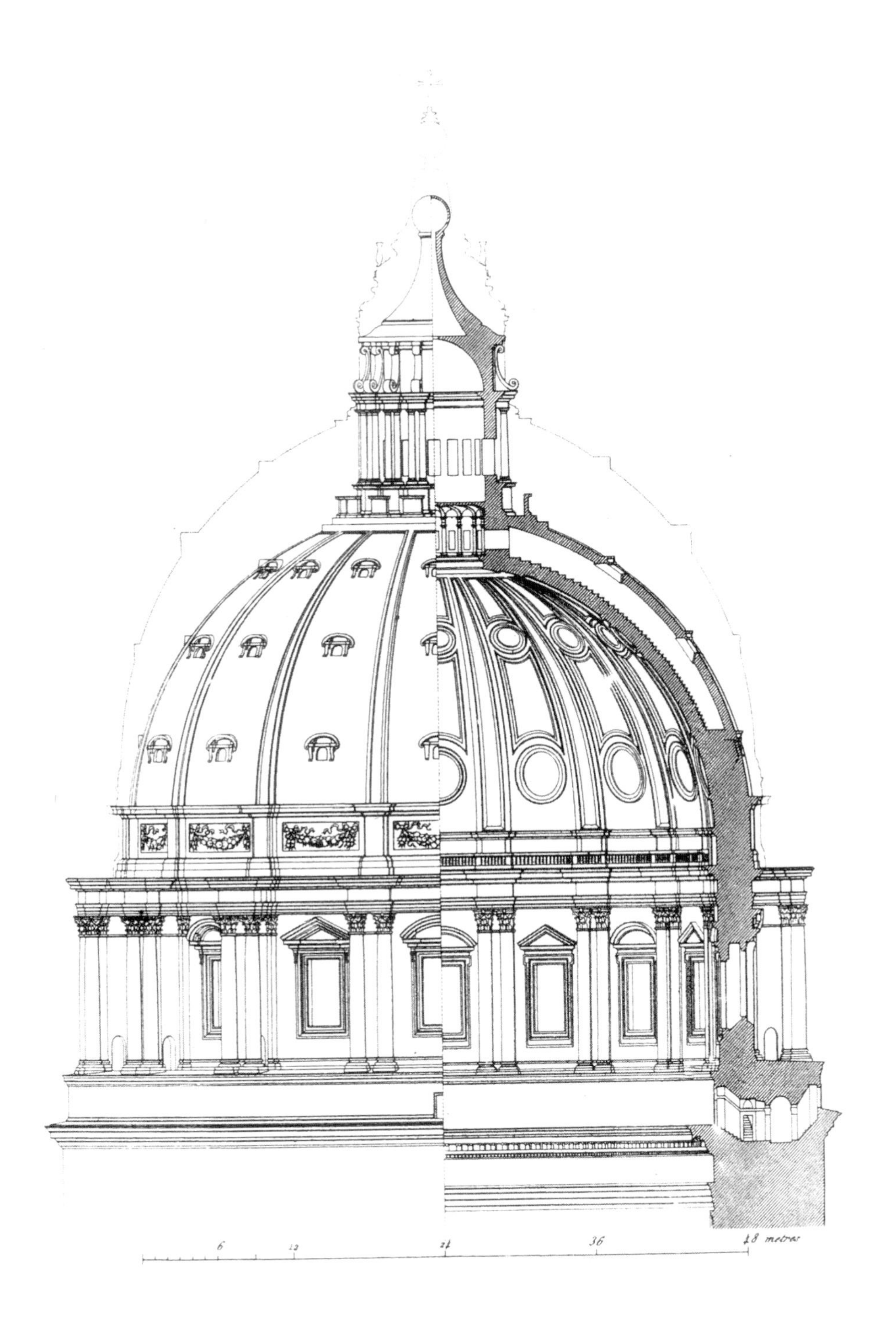

4.35 *St. Peter's Basilica. Michelangelo design, 1569. Cross-section through the great dome with outline of dome as built (Letarouilly).*

designation of Michelangelo Buonarroti (1475–1564) as architect, the worthiness of Bramante's centralized design was again recognized. Michelangelo created a bold, simple volume for the body of the church that was to be covered by a simple hemispherical dome articulated with prominently projecting ribs.

Unlike Bramante, Michelangelo understood that the brute-force structural solution of the Pantheon-like dome with its 6-meter-thick supporting walls was inappropriate for a lofty dome perched high above the crossing of the new basilica. Instead of the Pantheon, he studied the dome of the cathedral in his native Florence, for which he had once served as supervising architect (clerk of the works). His design shares with the dome of Florence the ribbed, double-shell construction (figure 4.35). Instead of its octagonal plan, however, Michelangelo's plan is circular, using sixteen ribs to tie together the inner and outer shells. Apparently he gave a higher priority to the architectural *form* of the hemisphere than to the possible structural gain to be had by pointing the dome. Nonetheless, he was not insensitive to structural concerns, as shown by a series of engravings published in 1569 by Dupérac that portray Michelangelo's concept of the buttresses of the drum as continuations of the deep ribs of the dome. Finite-element modeling of Michelangelo's dome design suggests that the drum-buttresses with their paired columns were too thin to offer significant resistance to the spreading of the dome (Robison, 256). The modeling also indicates that it probably would have been possible to counter hoop tension within the dome by placing a practical amount of iron reinforcement around the haunches.

Michelangelo, already in his sixties when appointed as architect of St. Peter's, realized that he would not live to see the project completed, so he had a wooden model built between 1558–1561 to act as a definitive guide. The change of popes brought about a suspension of work for the twenty-four-year period after his death, and when work on the basilica was resumed by Giacomo della Porta (1533–1602), the dome was given a more pointed profile, mainly to increase its visibility from the piazza in front (figure 4.36). Moreover, della Porta and his associate, Domenico Fontana (1543–1607), relied on iron "chains" (actually long, hand-wrought iron eye-bars) to counter the spreading tendency within the dome that was constructed between 1588 and 1593 (figure 4.37). Given continuing concern over the stability of the main piers and the foundations, della Porta and Fontana attempted to lighten the structure: both shells taper considerably as they approach the crown, and they follow the precedent of the Florentine dome by employing a thin outer covering, providing primary support for the heavy central lantern from the inner shell.

Finite-element modeling of the unreinforced della Porta dome has indicated low levels of tensile hoop stress at the dome haunches (Robison, 256). Apparently this tension was anticipated by della Porta and Fontana, who placed two iron "chains" (measuring from 6 to 8 cm square) about the base of the dome. Yet almost immediately after the dome was completed, cracks observed in a main supporting arch of the crossing had to be stitched together with iron cramps and plastered over. But by 1630, more cracks were seen to propagate into the dome, causing fresh alarm (Di Stefano, 14). And by the early decades of the eighteenth century, prop-

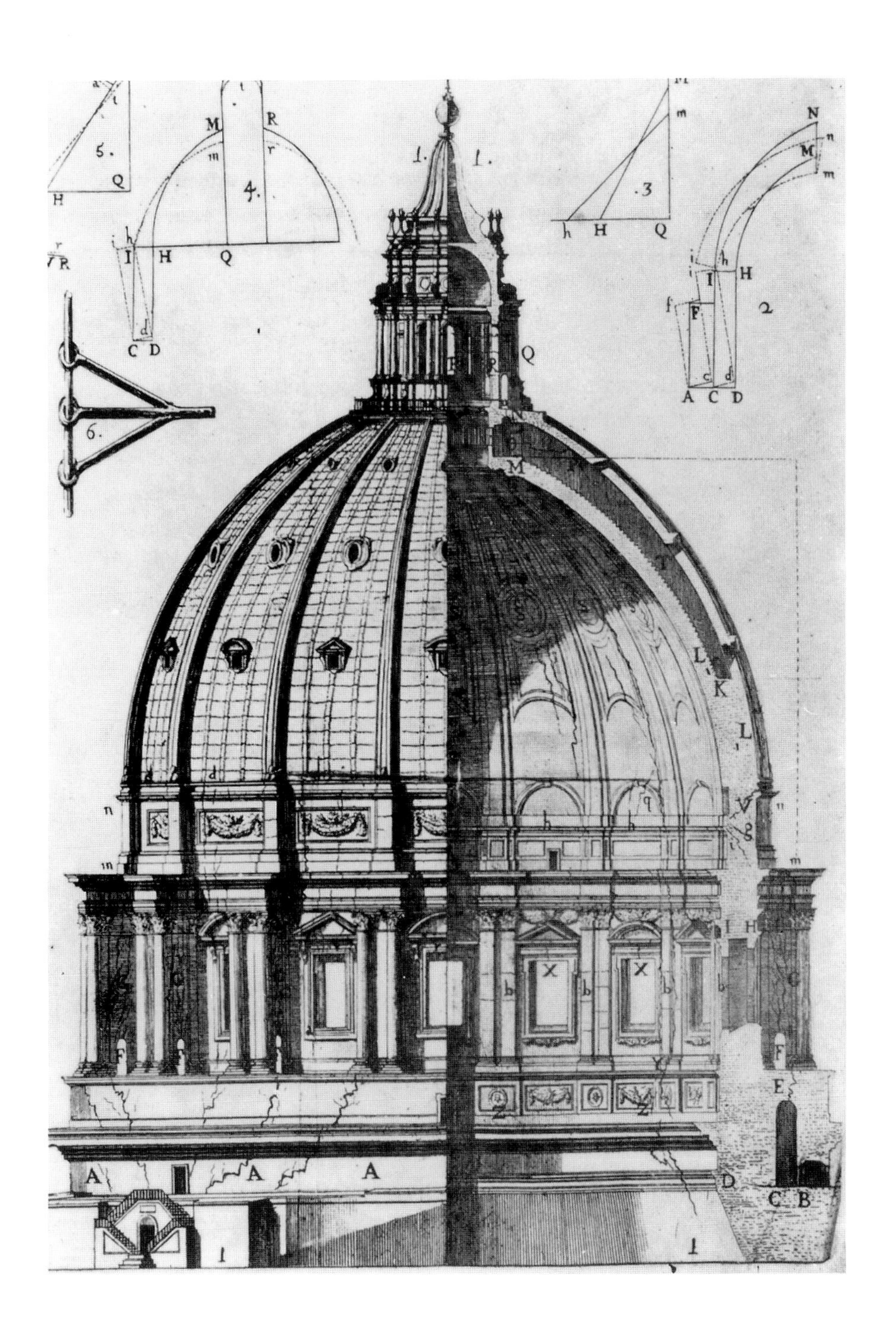

4.36 *St. Peter's Basilica. della Porta and Fontana design, ca. 1585. Cross-section through the great dome, showing crack pattern (Poleni).*

4.37 *St. Peter's Basilica. The great dome, 1588–1593.*

agation of the cracks had proceeded to the point where collapse seemed imminent, prompting the pope in 1742 to appoint Giovanni Poleni, a Venetian scientist, to recommend a solution. Following his detailed technical study, Poleni concluded that five additional iron chains of greater strength than the originals would remedy the problem (Poleni). While placing these, Poleni discovered that the original chains had broken, and he subsequently repaired them as well.

With Poleni's reinforcement, the dome of St. Peter's has remained stable since the eighteenth century. The buttresses of the drum and the corridor under the drum have required repeated maintenance, but the dome itself has experienced little movement. Only one thin vertical crack in the sector facing the Sistine Chapel is now evident. If the stability of this great dome, adequately reinforced with iron chains, could have had been predicted in the seventeenth century, the course of future large-dome design may have taken a different turn. Yet from the viewpoint of technology, it is fortunate that other means seemed necessary to solve the structural problems of large domes.

When called upon to recommend reconstruction for Old St. Paul's Cathedral in London, just a few months before it was destroyed by the Great Fire of London in 1666, Christopher Wren proposed to replace the central tower with a timber spire over a lightweight high outer dome constructed of timber and supported by an inner masonry shell (figure 4.38). The concept of this type of light, domical structure may have come directly from the much smaller dome of the Church of the Sorbonne in Paris (figure 4.39) built some thirty years before by Jacques Lemercier (ca. 1582–1654) and observed by Wren in 1665 on his only known trip abroad. In

4.38 *Wren proposal for restoration of Old St. Paul's Cathedral, London, 1666: cross-section through the crossing dome.*

4.39 *Church of the Sorbonne, Paris, 1635: cross-section (Marot).*

fact, this type of double-shell partial-timber construction can be traced from the Sorbonne church through a sequence of buildings, including S. Maria della Salute in Venice, designed in 1630 by Baldassare Longhena, back to medieval Venice where, as we have seen, it was used in the thirteenth-century enlargement of the high domes of St. Mark's Cathedral (cf. figure 4.17).

The starting point for the structural design of Wren's dome for the new St Paul's Cathedral was the double-shell scheme he had proposed for the old cathedral. Although its 31-meter diameter was to be only three-quarters that of St. Peter's, he seems to have been alarmed by reports of the problems of the Roman basilica. His close friend John Evelyn (1620–1706) had examined St. Peter's, and it was he who probably provided Wren with a first-hand account of its cracking. Evelyn's report must have contributed to Wren's hesitation in making a final choice from among the many dome schemes that he worked on almost until the actual construction of the dome, undertaken between 1705 and 1708.

Wren also had to face problems of another kind. Just as Bramante had done at St. Peter's, Wren began construction of the crossing piers well before the dome design was fixed. But instead of the four massive piers of the Roman basilica, Wren placed eight piers in line with the piers of the arcades and then added four bastions (the heavy tower-like structures at the reentrant corners of the transepts) to provide further stability. All of these structures were then joined together by an array of arches and barrel vaulting. Building records indicate that problems with differential settlement of the central piers were

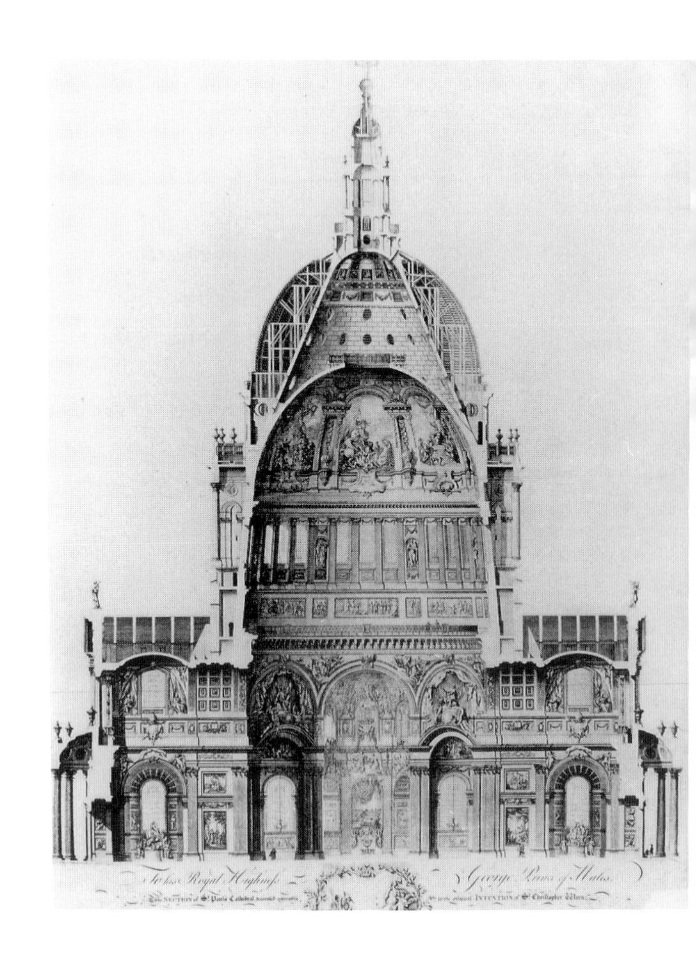

4.40 *St. Paul's Cathedral: cross-section through the great dome (J. Gwyn, 1753).*

observed from 1696 on (Fürst, 114). And it was these concerns that may well have prompted Wren to adopt in the end an extremely lightweight design (figures 3.77 and 4.40), a truly elegant solution both artistically and technologically.

Wren's design scheme is based on a majestic, light outer dome profile of lead-sheathed timber supported by an unseen chain-girdled brick cone, only 46 cm thick, which also supports a stone lantern of some 850 tons, and a separate brick dome (also of 46 cm thickness). In contrast to the action of the dome of St. Peter's, the brick cone of St. Paul's, formed by straight line generators (i.e., having a linear profile), is compressed by the heavy central lantern. Hence the cone, which also provides almost all of the support for the outer, visible dome, experiences compression rather than the pernicious tension characteristic of heavy semispherical domes.

In further contrast to St. Peter's, Wren's single iron chain proved sufficient to maintain the integrity of his relatively light structure against outward thrusts. Finite-element modeling of the triple-dome configuration indicated that stresses within the supporting masonry are generally low both under gravity and wind forces and that the single chain is well placed to fulfill its role (Mark 1990, 164). There have been problems with the dome's supporting structure, as mentioned above, but these are unrelated to the structure of the dome itself. In fact, had Wren arrived at this final dome design at an earlier stage of the project, before beginning construction of the central supporting piers, he might well have perceived that the piers could have been lightened and that some of the distress resulting from their settlement would have been avoided.

The triumph of Wren's solution is borne out by the fact that St. Peter's was the last of the great domes to be constructed entirely of masonry. Wren's structural scheme for the dome of St. Paul's became the standard for all the large dome projects that followed well into the nineteenth century, including the similarly scaled iron domes over the Cathedral of St. Issac in Petersburg, 1818–1858, and the United States Capitol, 1856–1864.

Bibliography

Abraham, Pol. *Viollet-le-Duc et le rationalisme médiéval.* Paris, 1934.

Alberti, Leon Battista. *Della Pittura,* ed. Luigi Malle. Firenze, 1950.

Baldwin Smith, E. *The Dome.* Princeton, 1950.

Bony, Jean. *The English Decorated Style: Gothic Architecture Transformed, 1250–1350.* Ithaca, 1979.

Bony, Jean. *French Gothic Architecture of the 12th and 13th Centuries.* Berkeley, 1983.

Butler, L. "The Entablature of Hagia Sophia as a Key to the Chronology of its Design and Construction." In R. Mark and A. S. Cakmak, eds., *Hagia Sophia from the Age of Justinian to the Present.* New York, 1992, pp. 57–77.

Cicognara, L., A. Diedo, and G. Selva. *Le Fabbriche e i monumenti cospicue di Venezia.* Venice, 1838.

Coulton, J. J. *Greek Architects at Work: Problems of Structure and Design.* London, 1977.

Di Stefano, Roberto. *La Cupola di San Pietro,* 2d ed. Naples, 1980.

Dorn, Harold, and R. Mark. "The Architecture of Christopher Wren." *Scientific American,* vol. 245 (July 1981), pp. 160–173.

Fitchen, John. *The Construction of Gothic Cathedrals.* Oxford, 1961.

Fürst, Viktor. *The Architecture of Sir Christopher Wren.* London, 1956.

Leedy, Walter C., Jr. *Fan Vaulting, A Study of Form, Technology, and Meaning.* Santa Monica, 1980.

MacDonald, William L. *The Architecture of the Roman Empire,* I. New Haven, 1982.

MacDonald, William L. *The Pantheon: Design, Meaning and Progeny.* Cambridge MA, 1976.

Mainstone, Rowland J. "Brunelleschi's Dome." *Architectural Review,* CLXII (Sept. 1977), pp. 156–166.

Mainstone, Rowland J. *Developments in Structural Form.* Cambridge, MA, 1983.

Mainstone, Rowland J. *Hagia Sophia: Architecture, Structure and Liturgy of Justinian's Great Church.* New York, 1988.

Manetti, Antonio de Tuccio, *The Life of Brunelleschi,* Howard Saalman, ed. University Park, 1970.

Mark, Robert. *Experiments in Gothic Structure.* Cambridge, MA, 1982.

Mark, Robert. *Light, Wind, and Structure.* New York/Cambridge, MA, 1990.

Mark, Robert and P. Hutchinson. "On the Structure of the Roman Pantheon." *Art Bulletin,* LXVII (March 1986), pp. 22–34.

Middleton, John H. *The Remains of Ancient Rome,* 2 vols. London, 1892.

Panofsky, Erwin. *Abbot Suger: on the Abbey Church of St. Denis and its Art Treasures,* 2d ed. Princeton, 1979.

Pevsner, Nikolaus. *An Outline of European Architecture,* 7th ed. Harmondsworth, 1974.

Poleni, Giovanni. *Memorie istoriche della gran cupola del tempio vaticano, e de' danni di essa, e de' ristoramenti loro, divise in libri cinque.* Padova, 1748.

Procopius. *On Justinian's Buildings,* trans. H. B. Dewing. London, 1940.

Rapporto sulla situazione del complesso strutturale cupolabasamento della cattedrale di Santa Maria del Fiore in Firenze, Soprintendenza per i Beni Ambientali e Architettonici per le provincie di Firenze e Pistoia, March 1985.

Rivoira, G. T. *Roman Architecture and its Principles under the Empire,* trans. G. Rushford. Oxford, 1925.

Robison, Elwin C. "St. Peter's Dome: the Michelangelo and the della Porta designs." *Domes from Antiquity to the Present,* ed. I. Mungan. Istanbul, 1988, pp. 253–260.

Rondolet, J. *Traité théorique et practique de l'art de bâtir,* 3 vols. Paris, 1834.

Saalman, Howard. *Filippo Brunelleschi: The Cupola of Santa Maria del Fiore.* London, 1980.

Smith, E. Baldwin. *The Dome: A Study in the History of Ideas.* Princeton, 1950.

Terenzio, Alberto. "La Restauration du Panthéon de Roma." *La Conservation des Monuments d'Art & d'Historie.* Paris, 1934, pp. 280–285.

Van Nice, Robert L. *Saint Sophia in Istanbul: an Architectural Survey.* Washington, DC, vol. I, 1965; vol. II, 1980.

Viollet-le-Duc, E. E. *Dictionnaire raisonné de l'architecture française du XIe au XVIe siècle,* 10 vols. Paris, 1854–1868.

Vitruvius. *The Ten Books of Architecture,* trans. M. H. Morgan. New York, 1960.

Ward-Perkins, John. *Roman Imperial Architecture.* Harmondsworth/New York, 1981.

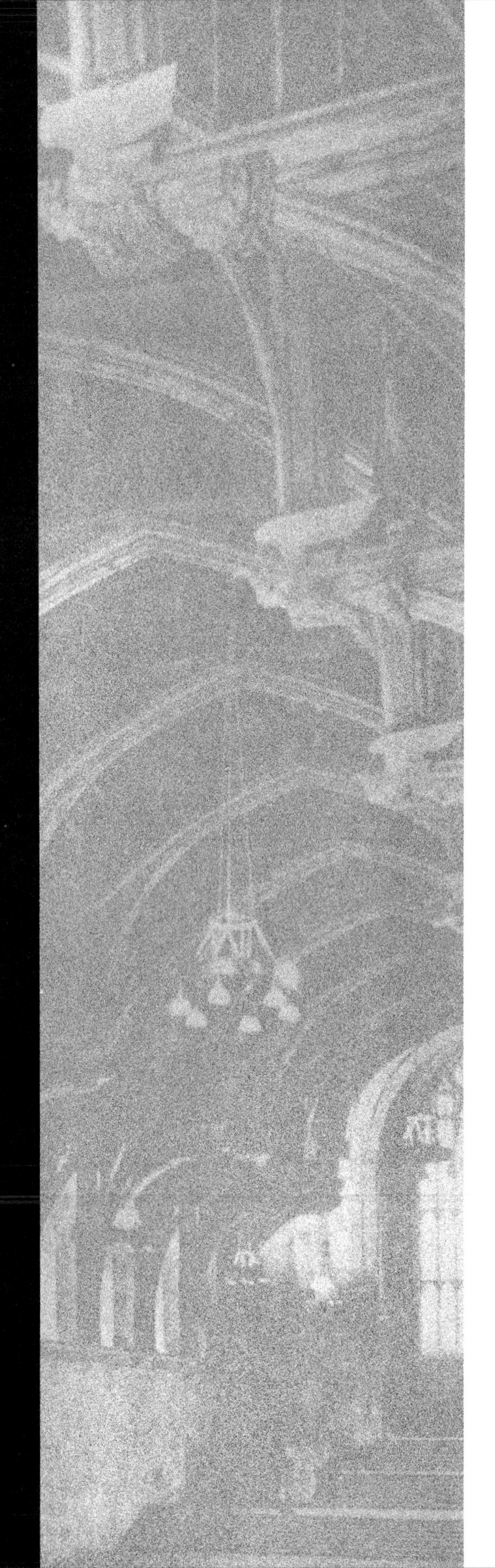

5 Timber Roofs and Spires

Timber roofs represent a wide variety of constructional forms and are fundamental to any technological appraisal of the evolution of European architecture. To build in stone, concrete, and brick, large quantities of timber were needed for permanent structures such as pilings, foundations, walls, and roofs, as well as for a variety of temporary supports including the machinery and hoists used during construction (see, for example, figures 3.17, 3.19 and 4.25). Yet apart from the growing field of vernacular architecture, the relationship between timber and masonry has been little appreciated in the general architectural literature. The present interest in technical aspects of premodern building and the current impact of timber-dating methods in a variety of disciplines are, however, changing the general assessment of the importance of structural carpentry.

Carpenter/engineers, who designed large-scale roofs, defenses, harbors, bridges, and lifting devices were craftsmen of high status, though much of their work has remained anonymous. Master carpenters shared responsibility with their mason colleagues for the ultimate success or failure of structures in wood and stone. They had to devise roofing systems that would insure stability against the forces of gravity and wind—the latter particularly crucial for the upper walls of tall structures such as Gothic cathedrals (see figure 3.4). In large-scale pitched roofs, this task was achieved by triangulated-braced structures having members connected by wooden-pegged joints.

Carpenters also had to select trees from the forests and supervise the hewing of timber for the job at hand. Hellenistic sources from Delos indicate that timber was precut to specified size (using a standard unit of measure, the cubit) before shipment (Martin, 38). After felling and hewing, the timber

5.1 *Gallo-Roman relief, Poitiers: hewing with a metal tool.*

could either be partially worked and shipped to the construction site or, to further reduce transportation costs, roofs could be pre-fabricated at full scale near the felling site with their members marked for reassembly, and then taken apart and transported either by carts or boats to the workers' yard at the construction site for final assembly.

Generally, from the classical era until the High Middle Ages, whole trees of appropriate cross-section were used, with a preference for straight hardwoods or the softer woods of the tall pines and firs. After ca. 1300, curved or elbowed oaks used in arched types of roof design gained in popularity, especially in multitiered roofs of the fourteenth to sixteenth centuries. How the timber was hewn—that is, whether the entire tree was roughly squared with an axe and smoothed with an adze, or split with axes and wedges, or sawn into halves, quarters, or planks—depended on the designated use of load-bearing members as well as traditional woodworking practices (figure 5.1).

By ca. 800 B.C., **tools** began to be made of forged iron and of low-carbon steel. These included the axes, chisels, burrs, planes, augers, and saws used by carpenters and also by masons (Martin, 38–46). Generally speaking, sawing either green or very hard, dry wood was a difficult process because of the tendency for the saw teeth to clog, wherefore the Hellenistic author and scientist Theophrastus recommends that the teeth be set at alternate angles to free them of sawdust (Theophrastus, V. 5. 3). The Romans, who expanded the technology of monumental roof carpentry, tempered steel to the desired hardness and workability for cutting tools (Forbes, 33–37). Steel-edged blades, used widely in the Middle Ages, help to explain the success of hardwood carpentry and the technical development of complex joints.

Empirical understanding of the capabilities of timber, its proper cutting, and the managment of **timber resources** are important corollaries to building with a material having variable properties. Factors such as availability, transport, and forest regulation were significant parameters within which patrons and master craftsmen operated. Timber, defined broadly by size, is constructional wood from a tree at least 60 cm in girth, which yields a heartwood of at least 15 cm square (Rackham, 23). Yet only certain species grown under specific conditions proved over time to be reliable for major construction. A general appreciation of the traditional knowledge about woods and the concern for quality timber helps to place both written and archaeological sources in a technological perspective. In fact, since the survival of roof timbers is so rare (one of the earliest published examples being the Casa del Telaio at Herculaneum; Trevor-Hodge 1960, 57), treatises and inscriptions constitute a major source of technical information for the ancient period.

Woodland resources can be broadly described geographically and hence climatically by distinguishing the ecology of the Mediterranean basin from that of northern Europe. Extensive forests of pine and conifers prevailed in Macedonia, Spain, Sicily (Mt. Etna), and central Italy—that is, the Greco-Roman civilization south of the Alps—while oak suitable for construction predominated north of the Alps and in the British Isles.

In the ancient Mediterranean region there was an enormous supply and wide variety of trees suitable for construction and shipbuilding, the primary woods being mountain fir, Aleppo pine, oak, and the preferred cedars and cypress of Lebanon and Crete. And although we have only scant direct evidence in inscriptions from carpenters that reveal timber specifications and costs, such as the uneven record from the Island of Delos (a cosmopolitan port for ancient maritime trade), we can obtain an idea of the dimensions of timber used in construction from them (*Inscriptiones Graecae,* cited in Martin, 33–35). Also, a growing body of information about the use of certain woods is being furnished by the expanding field of dendrochronology (figure 5.2).

One of the most important and influential ancient sources on timber is the work of Hellenistic scientist and writer *Theophrastus* (370–ca. 285 B.C.), a student of both Plato and Aristotle who wrote an extensive botanical treatise, *Inquiry into Plants,* which includes trees and timber for building purposes. In Book Five, he offers considerable practical information gleaned from ancient carpenters and woodsmen as well as botanists accessible to him at the Macedonian court of Alexander the Great. To the reports of others, he adds his own observations as well as a system of classification, all of which make his text the most original single document for the study of ancient carpentry until the end of the Middle Ages.

For roofs, Theophrastus finds that "Silver fir and fir are the most useful trees and in the greatest variety of ways, and their timber is . . . the largest." He observes, as is accepted still, that trees that grow on north-facing slopes produce bigger, more erect, straighter grain and "tougher" timber. The hardest woods available are several species of oak, which the carpenters wet before they attempt to bore holes, an observation that implies the use of dowels or pegs for connections of posts and beams (Theophrastus, V. 3. 3). Glue is also mentioned (and it appears in inscriptions from Delos as well). It is unlikely that glue would have been used for major structural joining without the addition of pegs or nails; it was probably employed more often in ceiling construction. Unfortunately, in contrast to the variety of iron clamps, bars, and dowels, whose traces are abundantly preserved in masonry (see figure 3.30), we know little about actual joints used in roofs by Greek structural carpenters (Ginouvès and Martin, 88–93).

Theophrastus, like Vitruvius later, is careful to inform his readers about when trees should be felled, the nature of their grain, and how they should be hewn, etc. Significantly, a section of his text is devoted to the question of load-bearing capacity: on the whole, silver fir is viewed as the best all-purpose timber for the carpenter. It is considered stronger because it does not split as do oak and olive (Theophrastus, V. 1. 5–8). Modern experience, however, demonstrates that timber quality (for example, straight grain, lack of knots and checks) is as important a determinant of strength as species.

Writing several centuries later, *Vitruvius* (ca. A.D. 25) also discusses a variety of trees for building, including oak and fir, which unlike the

5.2 Dendrochronology.

Dating of timber based on detailed measurement of variation in width and character of the annual ring patterns of a large sample of trees in a given region. Since tree rings reflect seasonal changes, only trees from temperate and arid regions have annual rings. Because oak is so long-lived and its outer sapwood is easily distinguishable, it provides ideal samples for study, although some conifers have been successfully dated. As in any statistically based method, the number of securely dated samples and the amount of ring-width data on file augments the ease with which a match with a master chronology can be made as well as its reliability. Climatic differences dictate that master chronologies be established for individual regions; about a dozen core samples are normally taken to date each phase of construction when such a chronology is used for dating a building in that region.

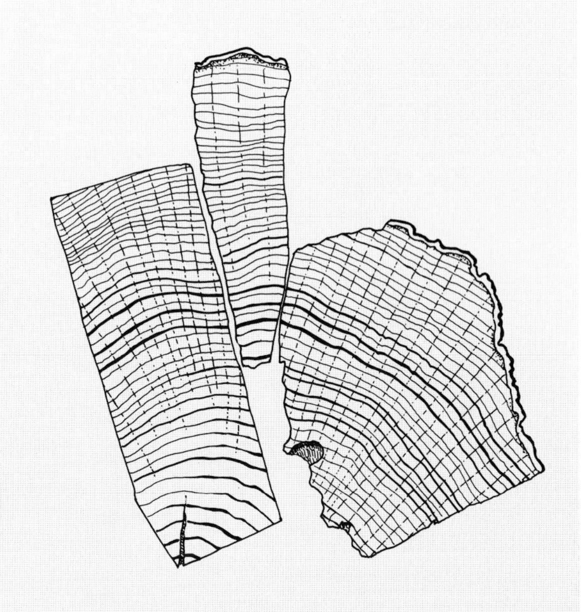

Beginning with existing, living trees of a particular species, such as oak or pine, sample ring widths are averaged to establish a mean sequence (i.e., a master chronology*). Undated specimens from the same tree type and region are then compared ring by ring to the known, dated sequence. When an undated sample of rings is matched with a master chronology, each of its rings can immediately be assigned to a calendar year. And if the bark and the last annual ring is preserved, the precise year in which the tree was felled can be determined. If the earliest ring of the undated sample predates the first ring of the master chronology, the cross-dated sequence then forms a "bridge" and extends the span of the chronology back in time. The key to establishing a master chronology spanning millennia involves joining ancient and prehistoric "floating" chronologies to a modern chronology based on trees of known felling dates. Once synchronization is found by cross-matching rings, even prehistoric sites can be fixed in real time. Recently, Hollstein (1980) has published a master oak chronology for northern central Europe extending to 700 B.C. This is one of a number of important, European "master chronologies" developed for archaeological dating in the last decade (cf. Baillie 1982). Subsequently, Kuniholm and Striker (1987), working south of the Alps, have established an Adriatic or "North Greek oak master chronology." With the establishment of an ever-increasing number of regional master chronologies, tree-ring analysis has evolved as a primary tool for dating early buildings; as a by-product of this research, we know much more about the regional species of wood used in historic carpentry.*

5.3 *Trajan's Column, Rome, Roman soldiers felling oak trees with* axes.

more costly cedar and cypress are subject to decay. In addition to treatises such as those by Vitruvius and Theophrastus, literary descriptions of buildings and of woodland practice by ancient authors including Julius Caesar, Strabo, Dio Cassius, and the Elder Pliny give a good indication of the vast spans (often exceeding 25 meters) of Roman roofs and bridges.

North of the Alps, the preeminent woods for any major roof construction were oak and chestnut, which for structural purposes are practically the same. This appears to have been the case from ancient times (i.e., the Roman occupation of Britain and Gaul) until the modern era, with the exception of the extensive importation of Baltic pine and fir as well as oak boards to Britain in the Middle Ages (Salzman, 246–247). The forest scenes on the *Column of Trajan,* depicting the Dacian campaigns of the early second century A.D., are perhaps the best visual record of ancient tools and timber construction. The Roman army on the Danube frontier was clearly occupied with hewing oak and building fortifications with trussed timber towers and rampart walls (figures 5.3 and 5.4).

Building timbers were to a certain extent *produced* in the forest in order to provide a desired shape (straight, or the later curved blades) and size. While global environmental protection and scientific ecology are modern developments, documents of the period indicate that in the Middle Ages woodlands were highly regarded and legally protected not only as hunting preserves for the nobility, but as precious resources that provided timber. Premodern societies, at least in England, did not treat their forests with reckless abuse in order to build merchant ships and navies. Even during the extreme demand for timber

5.4 *Trajan's Column, panel detail: Roman towers and ramparts; soldiers hewing timber.*

during the Napoleonic wars, when the largest recorded number of trees was felled from the main forest preserves, there was still no critical shortage in Britain that would endanger the forest from regenerating itself within a generation or so (Rackham, 100). Yet if one takes seriously Abbot Suger's well-known and rather theatrical account in *De consecratione* of his arduous search for the twelve great beams needed for the roof of St. Denis (Panofsky, 95–96), it would appear that large trees (no doubt oak or chestnut) were less available in northern France, perhaps owing to the great clearings and expansion of arable land in the twelfth century. Nonetheless, the oak tie-beams of the roof of Notre Dame in Paris, probably dating from the late twelfth century, are 24 × 25 cm in section and over 14 meters in length.

The interrelationship between forest resources and actual forms of timber construction is reflected, too, in the general shift away from the large-scaled, *aisled construction* (post and beam construction) typical of medieval halls and barns to aisleless structures or those in which the *bay interval* between supports is increased. By reducing or eliminating the series of tall, heavy load-bearing posts of the central aisle, for example, trees of lesser length and section, which were easier and faster to fell and transport, could be used for roof trusses. Indeed, the scarcity of long, straight timbers is underlined by the abundant evidence after ca. 1200 of their reuse.

Apart from whatever visual appeal they possess, all roofs have two functional components: (1) the internal structure, an assembly of rafters, cross beams, struts, etc., that support the roofing

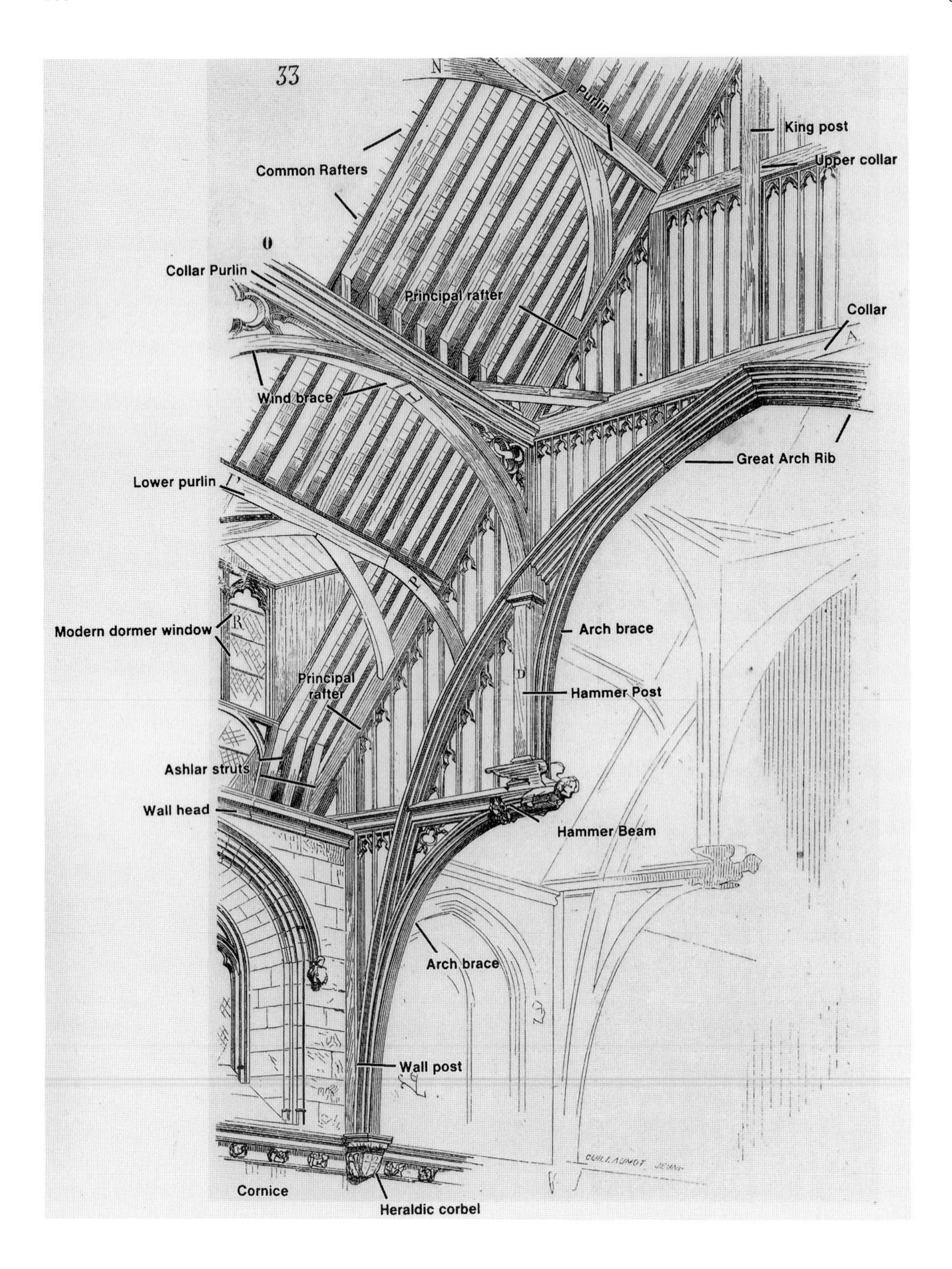

5.5 *Westminster Hall, London, 1393–1397: roof framing (after Viollet-le-Duc).*

material; and (2) the outer roof covering (cladding): tile, slate, stone, bronze, lead, or for buildings of lesser social status, turf, clay, thatch, and reeds, that provides protection. The type of **roof covering** may influence a roof's pitch, if it has not already been predetermined by structural or aesthetic tradition. For most major buildings, though, the variables of roof design, covering, and the structural/aesthetic relation to the building as a whole interact simultaneously. Lead sheets nailed onto boarding, for example, the preferred material for steep roofs in northern Europe after ca. 1180, could be used at virtually any pitch, whereas thick slate or flagstones were traditionally laid at about 35 degrees from the horizontal (Brunskill, 60), although recent evidence indicates that slate was commonly used on steeper pitched roofs in England (Crook, 143). Also, certain roof coverings tended to become traditional in particular regions naturally rich in clay or slate, such as the continued use of half-round Roman tiles in vernacular and ecclesiastical buildings of southern France in contrast to the predominance in northern Europe of thatch, slate and, for elite buildings, lead (see Meirion-Jones, 3–5).

Two basic roof types encompass most structural carpentry: (1) roofs whose internal timber structure is meant to be seen from below, termed *open-timber roofs,* as found in some Roman basilicas, Anglo-Norman churches, medieval Italian churches, and especially late-medieval English halls (e.g., Westminster Hall, London, figure 5.5); and (2) *concealed roofs,* whose timberwork is intentionally hidden. These may include the cellas of Greek temples, where coffered ceilings and stone pediments traditionally hid the structural timbers, as well as the major category of roofs built above the vaults or ceilings in medieval churches.

It is important to realize that early roofing systems are not trusses as defined in modern engineering usage, even though this term occurs widely in the literature to describe almost any major transverse support for longitudinal members of a roof. A true **truss** is an open assembly of linear members linked together to form a triangle or combination of triangles, the only polygon whose shape cannot be altered without changing the length of its sides. Because a triangle cannot be distorted, as could for example a square deformed to generate a rhombus, a truss is *not* dependent on rigid joints for stability. Most important, member stresses are lower than in an equivalent beam because the major forces in truss members are direct tension or compression rather than bending, as illustrated in figure 5.6.

Employing the principle of the truss, that is, tying the feet of the rafters, gained early builders considerable advantages: the ability, via a system of triangulation, to use smaller timbers (of lesser length and especially of smaller cross-section) to create wide spans. By controlling the tendency of the rafter feet to spread outward, the horizontal, overturning force at the wall head was restrained. The implications of this technological development, whose origins are obscure, were significant not just from the perspective of creating larger spans above less massive walls, but also aesthetically. With the capability of opening up the top of the wall without compromising stability, efficient base-tied trusses allowed light to enter the building interior at a high level without large amounts of buttressing (for example, in the Roman basilicas of Old St. Peter's, and St. Paul's Outside the Walls, as illustrated in figure 5.7). Apart from trusses, a second term frequently encountered in the English literature on early carpentry is *timber framing,* which refers to wall-framing rather than roofs

5.6 Force distributions in apex-loaded simple roof trusses. *With loads applied at the member joints, internal forces are direct compression and tension. Nominal magnitudes of member stresses are then simply found by dividing the member force by its cross-sectional area (see figure 3.9). Note that a slender, horizontal tie beam cannot accept any significant* vertical *force near its center. But with appropriate joining, the vertical member can serve as a tensile "hanger," supporting the center of a long tie, particularly if the tie is spliced. Note also the reduction of forces in members of the taller truss which, in turn, also serves to reduce member stresses.*

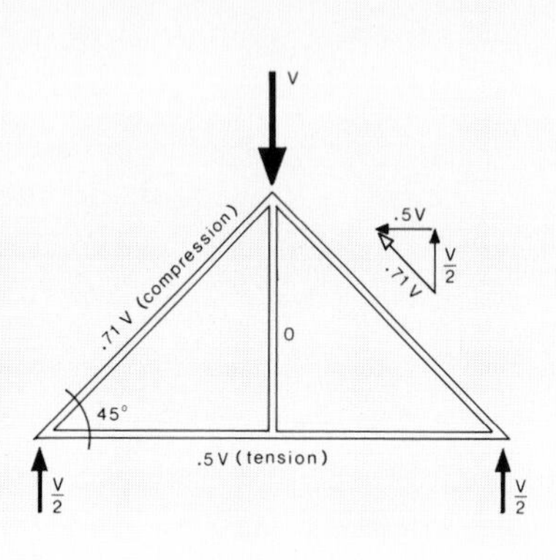

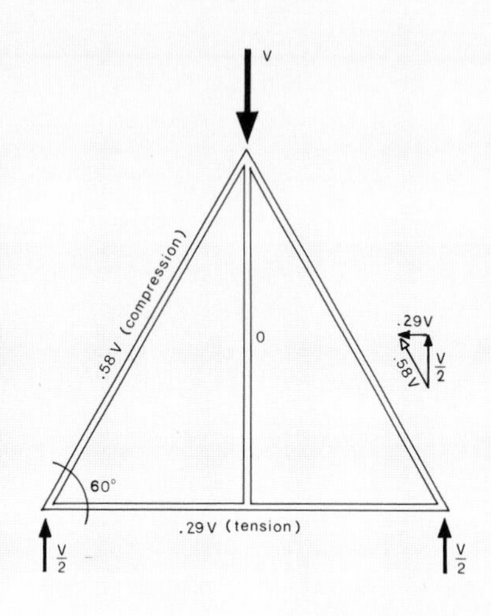

per se. However, in aisled construction, timber posts and lintels form a framed, *bay system* of construction within masonry walls, as for example in the aisled barn of Parçay-Meslay (figure 5.8), which is representative of a large category of monastic, aisled barns (Horn).

In the earliest roofing systems, posts, beams, and *purlins* (longitudinal supporting members generally parallel to the roof's ridge) were held in place by simple propping or by direct abutment from a fixed masonry support, a system that correlates closely with post and lintel construction discussed in chapter 3. As spans increased and the demands on traditional methods of roofing became more complex, a means of connecting timbers was needed so that forces could be transmitted appropriately through one member to another to an ultimate support. There are, generally speaking, four critical areas of **joining** in large-scale framing, as illustrated in figure 5.9: at the *apex* where the rafters connect, especially where a central post or hanger rises to the roof peak or ridge piece; at the *rafter foot,* where the tie beam and rafter meet above the wall head; the region of the tension splices (called *scarfing*) in the tie beam, particularly for long-span roofs; and at the ends of a hanger (or hangers) that support the spliced tie beam. From the connections of the members, their comparative cross-sections, and the angle at which they are cut, one can usually deduce what the original designer intended, that is, his understanding of the nature of the forces (compression, tension, or shear) transmitted from one member to another. For this reason, with timber joints we are far more in touch with the structural perceptions of ancient builders than in other modes of construction.

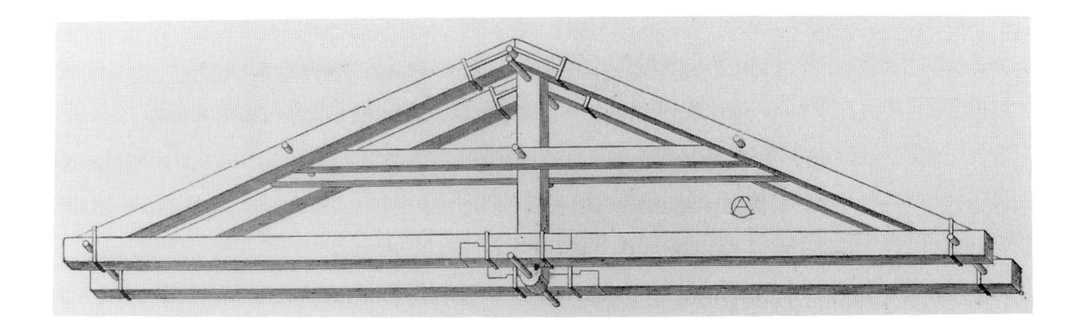

5.7 *St. Paul's Outside the Walls, Rome, late-fourth-century roof truss over the nave repaired in* A.D. *816 (Rondelet).*

5.8 *Parçay-Meslay: aisled grange barn (1211–1227), roof rebuilt, late fifteenth century.*

5.9 Critical areas of joining in a roof truss. *(a) Apex: simple lap joint with face peg. (b) Rafter foot: half lap and dovetail joint. (c) Tie beam splice: splayed and tabled scarf joint. (d) Hanger for tie beam: notched lap joint.*

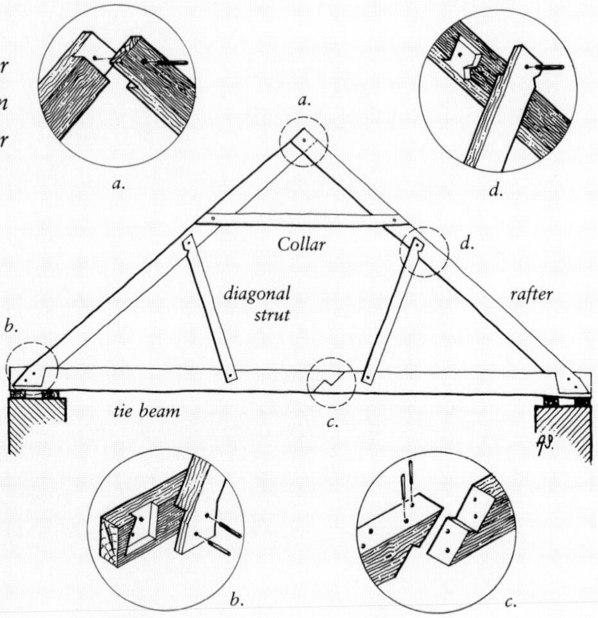

The common methods of joining, which were later enhanced and refined, derive from a few simple procedures. One of the most fundamental is *overlapping,* in which one member is passed over another at a right or an oblique angle and secured by a face peg, with neither timber being cut. A more secure and widely used version of overlapping, *half lapping,* is obtained when about half of each timber is cut across the grain to form a trench or slot to fit onto the other timber. Although it weakens the individual members by removal of a substantial amount of section, the joint, when pegged, is quite secure and was used ubiquitously for centuries, especially for joining rafter couples at a roof apex, as illustrated in figure 5.9.

To create a joint to accept tension, the simple, open half-lapped joint was modified by splaying and/or notching the timbers (as also illustrated in figure 5.9) to form the *notched lap joint,* or the *dovetail lapped joint,* both of which when tight and prevented from shrinking would resist withdrawal. A third procedure deriving from the need to create longer members from shorter timbers is *scarfing,* a common method involving splaying the ends of two timbers along the grain, overlapping them, and securing the long joint by a series of wooden pegs. To help lock the two pieces into position, carpenters generally cut a break or "table" into the splay and used pegs, wedges, or keys to tighten the assembly. Lengthening joints may be quite complex, but the so-called *splayed and tabled scarf joint* used for the tie beam in figure 5.9 provides a simple example. In the case of a scarf joint in a large cross-section beam subject to bending from its own weight, it may be necessary to reinforce the juncture with a clasp or *hanging joint,* originally of wood and later by an iron strap.

5.10 *St. Vincent of Soignies: nave roof framing, eleventh century (axial plate is not original).*

The fourth and certainly most versatile joint type involves connecting timbers mainly in compression by inserting one into another by *mortise and tenon.* The simplest version is a so-called short stub-tenon joint, in which the tenon (projection) fits into an appropriately sized mortise without a peg—a joint well suited to a vertical post in compression. More elaborate forms of this joint involving an oblique angle of entry require shoulders for the tenons to fit snugly, as well as a peg or pin through the sides of the mortise and the tenon. The variations of this ancient joining method are numerous, and several will be mentioned when pertinent to the discussion of structural behavior.

In all roofs, an essential distinction must be made as to how the load of the roof is carried and distributed to the supports below. In some **roofing systems** all of the rafters or trusses may be of the same design, size, and weight; i.e., all the major members have approximately the same cross-section "scantling") and are generally closely spaced. This system of roof carpentry, known as *uniform scantling,* and forming a widely prevailing roof type in the Middle Ages, is analogous to a barrel vault in its continuous support requirements. The repetition of closely spaced trusses, usually a meter or less apart—the normal system for early medieval roofs (figure 5.10)—thus distributes uniform loads along the length of the supporting wall, and the roof covering, without bays or purlins, provides the lengthwise stiffening (Smith 1958, 113–118).

In a different system, comparable to the support conditions for groin vaults discussed in chapter 4, the major load of the roof is concentrated at wider bay intervals by the use of heavy transverse supports called *principal trusses.* This *bay system* of roofing uses heavier, so-called *principal rafters* that

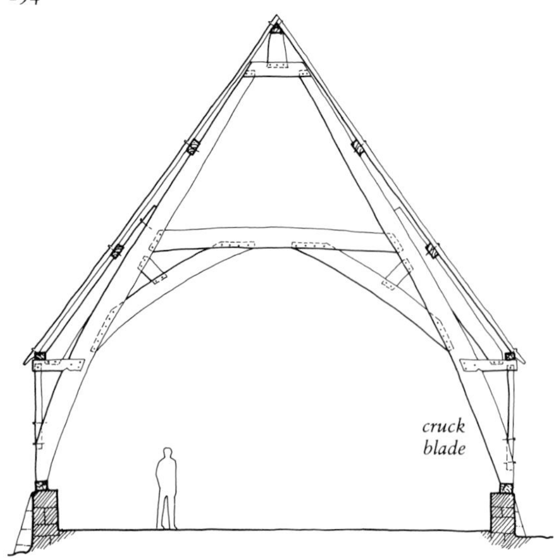

5.11 *Leigh Court Barn, ca. 1325: largest-span (10.2 m) extant full cruck (after F. W. B. Charles).*

support lengthwise members called *purlins*. The purlins, in turn, provide support to lighter *common rafters* (as illustrated in figure 5.5). By using the system of heavy principals, the carpenter can concentrate the load at intervals related to specific support conditions, such as the spacing of the outer colonnade in the Greek stoa, or on the solid walls alternating with clerestory windows as in Westminster Hall.

Within the category of principal-rafter roofs with bays and longitudinal purlins, a notable roof type is the *cruck frame,* defined simply as two inclined timber blades (usually bent or curved) extending from near ground level and joined at their apex to combine roof and wall into one integral load-bearing structure (figure 5.11), not unlike a modern three-pinned rigid frame of steel or reinforced concrete. (And like their modern counterparts, they were commonly raised as a unit rather than assembled in stages.) Full crucks and cruck-like structures involving curved *short principals* (whose origins and forms are still debated among specialists; Alcock, 1ff.) are of primary importance in vernacular architecture and also in combination with other roof forms developed in the later Middle Ages in wide-span multilevel construction.

Ancient

Ancient Mediterranean roofs were generally low in pitch, usually about 20 degrees (figure 5.12). The earliest roofs were covered by reeds, mud, clay, and thatch, but after the seventh century B.C., Greek and Roman roofs were traditionally covered with terra cotta tiles of varying design (Coulton 1977, 32–35). The use of clay tiles in classical Greece thus reinforced the tradition of low-pitched roofs, since tiles were not usually fastened onto the wooden battens but were held in place by gravity, or occasionally, a bedding of clay supported on what might be termed *common purlins,* close-set longitudinal timbers parallel to the ridge (Martin, 46–47).

The structural carpentry that supported the outer covering, normally with at least some degree of pitch for drainage, was based on the post and lintel system of vertical supports (*props* in compression) and horizontal or inclined cross-beams (*lintels* in bending). Evidence for three types of roof survives: (1) the terrace roof, which is nearly flat; (2) the shed or pent roof with one slope; and (3) the two-sloped ridge roof. From the Archaic period on, nearly all roofs of major buildings were of the double-sloped, ridge and cross-beam type, thus marking a distinction from the flat roofs of the Bronze Age and domestic building practices of the Near East (Meiggs, 191, 222).

The third type, the tiled, ridge and cross-beam roof (see figure 5.12) used conspicuously for temples, treasuries, and stoas on the Greek mainland,

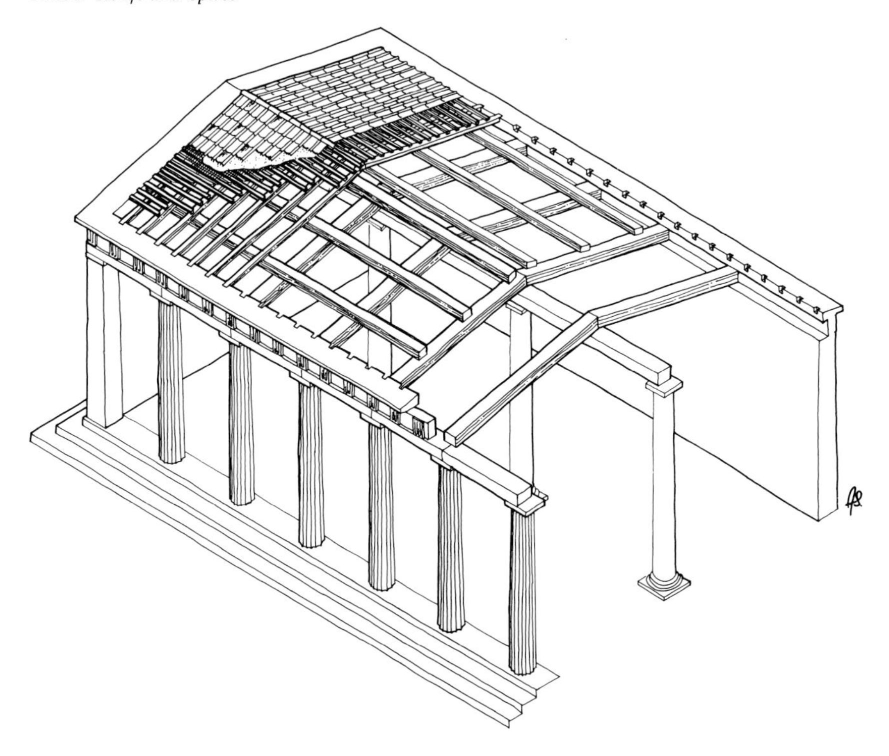

5.12 *Aisled Greek Stoa: cross-beam roof construction (partially after Coulton and Martin).*

underwent structural and aesthetic development from the Archaic to the Hellenistic period. From a technological perspective it is important to keep in mind that the Greek ridge and cross-beam roof with purlins was used for centuries before the triangulated truss, whose origin is discussed below. Hence the single spans that could be achieved in this manner were limited from the outset by the timber resources and the capability of a load-bearing beam in bending, the limit being around 12 to 13 meters, a figure based on the recorded spans of ancient structures (Martin, 33–36). In its most basic form, the ridge and cross-beam roof consisted only of sloping cross-beams and hence is termed a *simple rafter roof,* with the ridge supported by a masonry prop or column. The tiles rest directly on the beams, which serve as the principal rafters, and there is no superstructure of purlins, rafters, and battens under the roof tiles.

In the more highly developed *classical Greek roof,* however, a system of primary and secondary timbers evolved to span the wide porches and internal cellas of temples such as the Parthenon (figure 3.31). The major members in such roofs are: (1) the longitudinal *ridge piece* that supports the roof's apex (i.e., the upper ends of the main sloping rafters, which are generally very large-sectioned members). The ridge, in turn, is commonly supported by (2) a vertical timber prop resting on (3) a large-sectioned horizontal lintel (a bearer beam in bending) or on some kind of masonry (such as the inner colonnade illustrated in figure 5.12). The horizontal, timber lintels in a typical Greek temple are masked by the classical entablature and pediment so that from the exterior, there is one horizontal level at the base of the roof extending from the outer colonnade across the entrance porch. In reality, the timber roof is achieved by multiple spans, so that the outer, inner, and cella colonnades, for example, serve to support a series of lintels. The outer ends of these horizontal members are normally housed in masonry sockets in the cornice blocks, carried either by the peristyle or by an inner entablature of the cella. The sloping cross-beams, or principal rafters, carry the longitudinal purlins, which are parallel to the ridge piece and are frequently substantial timbers; for example, the six purlin sockets located on the interior of the west gable of the Parthenon measure 94 × 95 cm (Meiggs, 197).

The classical roofing system with multiple spans achieved by lintels, principal rafters, and common purlins has the advantage of allowing the architect more choice about the disposition of internal supports, especially if he had access to long, sturdy timbers. Structurally, however, ancient Greek roofs, despite their variety, still depend on traditional post and lintel construction, although major and minor timbers may be arranged in different ways depending on the resources at hand (Coulton 1977, 157).

The *secondary timbers* comprise the common rafters, battens, and whatever underlies the roofing material; here timber of lesser quality could be used. Masonry walls, columns, gables, purpose-built supports, and sockets in cornice blocks to house timbers were thus critical adjuncts to roof construction. Moreover, in classical Greek roofs timber joining is *not* an important technical factor, since there are no tension members, and the timbers in compression and bending are supported primarily by masonry and secondarily by other timbers used as props in compression. In fact, the only connections necessary for these timbers are: (1) the juncture of the ridge piece and the rafters and (2) the grid support for the tiles formed by battens etc., and possibly, (3) the use of a timber prop to support a purlin—all of which could be easily achieved by simple trenching (tongue and groove), dovetail, or mortise and tenon connection.

As discussed in chapter 3, the **Parthenon** marks a structural change in the organization of the interior of the traditional Doric temple. This innovation involves roofing because it concerns not just a general increase in building scale and design, but a dramatic increase in the span of the cella to almost 10 meters, the widest ever built in mainland Greece. And although the Parthenon was designed to appear to be entirely a work of marble, timberwork of considerable dimensions was required to handle the wider spans and to support the tiled gable roof and marble coffered ceiling below. This change in interior space is clearly delineated in the comparison of the Parthenon with its most important predecessor, the Temple of Zeus, Olympia, as summarized in chapter 3.

Since there are no remaining purlin sockets in the east gable of the Parthenon to correspond with those of the west gable, and since the west purlins are not aligned with the cella colonnade, it is reasonable to assume that the main roofing system did not continue over the cella itself (Stevens 1955, 250; cf. Trevor-Hodge, 47). Hence the cella must have been spanned independently using large beams supported by the cella walls and inner colonnade, spaced to accommodate the gigantesque statue of Athena Parthenos (figure 3.31). The conjectural framing of the east end of the Parthenon suggested by Stevens consists of a central ridge purlin (which would naturally correspond to the west gable) supported on a timber prop and three additional side purlins on either side of the ridge, supported by the cella colonnade, cella wall, and peristyle respectively, similar in fact to the section after Coulton (figure 3.31), but at a higher level in relationship to the statue of Athena. Modern calculations suggested that the depth of a square-sectioned beam required to carry the ridge loading would be about 65 cm (Stevens, 251). Iktinus and Kallikrates probably used the biggest timbers available, since the enormous size of the purlin sockets on the west gable suggest the use of very large-sectioned timbers overall—very likely cypress from Crete, a premium timber in both cost and quality (Meiggs, 200). One can speculate that the carpenter in charge may have intentionally reduced the size of the ridge piece and thereby decreased the cross-section of the prop required to support it in order to lessen the point load at the mid-span of the central cross-beam.

As it is generally assumed that Greek builders, at least those in Athens, did not use triangulated timber trusses, the wide span of the Parthenon cella presented a challenge, and it has been suggested that it might have been a potential candidate for a true truss rather than a long-span beam supporting a heavy ridge piece at its midpoint (Trevor-Hodge, 40–41). All the evidence, however, points to the continued use of the traditional "prop and lintel" system described earlier. The central lintel acts as a "standalone" structure and clearly not a tension tie. Indeed, as far as roofing is concerned, builders on mainland Greece understood that a beam of substantial scantling functions well if the central load of the ridge prop is not too great and the span is not excessive, that is, over 12 meters.

In contrast to the Parthenon, temples of Magna Graecia, and especially Sicily, as early as the sixth century B.C. have wide spans without an internal colonnade; for example, at the Temple of Zeus, Olympios, Akragas (Agrigento), where the cella spans practically 13 meters. On the basis of these unprecedented spans, without aisled construction, Trevor-Hodge argues that "the only logical conclusion to draw from this is that the Greeks never used the truss in [the sixth-fifth century B.C.] while the Sicilians did from . . . about 550 onwards" (Trevor-Hodge, 40–41). Yet archaeological evidence for this inference is tenuous. Moreover, excellent forest resources on Mt. Etna gave the Sicilians access to timber that probably could span as much as 13 meters *without* their necessarily devising a true truss. We are then left with the unresolved question as to how extensively the principle of the truss was known or used in the classical Greek world. At present, it is prudent to assume that the Sicilians had tall, sturdy trees that could provide timber for long beams and that the tension-tied truss evolved at a later date, probably in the Hellenistic period. But like the true arch also known to the Greeks, the truss was not extensively employed until the expansion of Roman architecture beginning in the second century B.C.

Imperial Rome

The architecture of the mature Empire, dominated by colossal vault and dome construction, called for the design of sturdy centering, scaffolding, and sophisticated formwork for complex curvilinear forms. As discussed in chapter 3, one can still discern the imprint of the wooden planking in many Roman structures. Roman carpenters were equally successful in achieving large spans with trussed timber roofs, particularly in the open-roofed basilica. Indeed, it should be stressed in this context that Roman carpenters, whether they actually invented it or not, were pioneers in developing stable timber trusses that used lighter-sectioned members (as opposed to the classical Greek prop and lintel construction).

5.13 *Trajan's Column, panel detail: Danube Bridge of Apollodorus of Damascus.*

The range of monumental Roman timberwork, even excluding shipbuilding, machinery, and military devices, is considerable and extends from simple turf and towered log forts of the provincial *limes* (as illustrated in figure 5.4), to enormous granaries, and bridges across the Rhine and Danube, as well as numerous wide-span, open-roofed public halls. Indeed, the trussed Danube Bridge of Apollodorus of Damascus (figure 5.13), Trajan's gifted engineer-architect, was a marvel in its time and reckoned as one of the emperor's greatest building achievements. Ranging over the swift current of the Danube in (today's) Romania, the bridge was constructed of twenty masonry piers each 70 Roman feet apart, 150 feet high, and 60 feet wide (Dio Cassius, 68: 13). Spanning the interval from pier to pier were a series of trussed segmental arches that sprang from a triangulated support, likely built of oak. This array of arches formed the support for the

5.14 *Trajan's Column, panel detail: timber-framed amphitheater.*

joists of the deck. As was typical of nearly all the timber structures depicted on Trajan's column such as turrets, ramparts, and amphitheaters (figure 5.14), the deck of the Danube Bridge was flanked by timber railings, shown in bird's eye perspective. The artist, who carved the marble panels and who likely worked from drawings or a sketchbook of sorts made at the site, had to simplify the structure for the sake of artistic clarity. For example, the radial timbers must represent clasps to secure the jointed timbers of the arch. The system, however, acts essentially in compression and recalls the Roman use of arches—only in this case the members are trussed timbers rather than masonry voussoirs. Unfortunately, this magnificent structure was demolished by the Emperor Hadrian, who had a profound personal dislike of Apollodorus.

Although much of our visual and archaeological evidence for monumental timberwork derives from such Roman military structures and provincial sites, our discussion will concentrate on the large open-timber roofs of Roman civic buildings, namely, **basilicas** (literally, royal buildings). It must be remembered, however, that the timber-roofed basilica of the Empire was already a building type of considerable antiquity, originating possibly as early as the second century B.C., when the Romans began to absorb the art and wealth of Hellenistic Greece and Asia Minor (Meiggs, 225).

The basilica that had such a profound influence in western architecture was essentially a long peristyle hall. The Roman version was frequently an aisled structure with a wide central aisle (the nave). Typically, the nave was taller than the aisles (see figure 3.43), and from the late Republican era on, its internal span was three to four times greater than the flanking aisles (Vitruvius, V. 1. 4). Whether the supporting walls of the nave incorporated the clerestory windows or it was lit by windows in galleries in the second story over the side aisles, the basilican nave presented a major roofing challenge to Roman builders. Relieving the upper walls of horizontal forces generated by the main-span roof covered with terra cotta tiles was critical to stability. For nave spans from about 13 to 30 meters (typical of Roman as opposed to Greek monumental scale), the triangulated tie-beam truss placed at bay intervals offered an ideal technological solution. For long spans, paired trusses incorporating hanging (tension) members clasped by double ties (as seen at St. Paul's Outside the Walls) provided support for the tie beam along its length, and these doubled trusses formed the principals to carry the roof's superstructure through purlins. Fortunately, firs from the higher altitudes of the Apennine forests were available for such long tie beams (Meiggs, 227).

Imperial roof carpentry of major civic buildings is difficult to reconstruct, because these structures have perished or have been altered continually over time, so that no actual Roman roof remains intact. Nonetheless, we have literary descriptions, coins, and archaeological evidence of enormous timber-spanned buildings beginning with pre-imperial structures like the basilica Aemilia of ca. 100 B.C. on the north side of the Forum, whose central span was about 17 meters and later, completed in 7 B.C., Agrippa's *Diribitorium* (the place where votes were counted) described by Dio Cassius as the largest hall (30½ meters wide) ever built under one roof. The tie beams, likely of Aleppo pine or silver fir, of the immense trusses (partially damaged in the fire of A.D. 80) were taken down and could not be replaced on the same scale at the end of the second century A.D. (Meiggs, 255).

Although the colossal single span of the *Diribitorium* was not duplicated or surpassed, the spans of other major Roman basilicas indicate the continued tradition of the wide-naved hall. The Basilica Ulpia in Trajan's Forum had a 26-meter span, and Old St. Peter's a 23-meter span for the nave. Also in Rome, the basilica of St. Paul's Outside the Walls (see figure 5.7), whose roof framing spanned 24⅓ meters, was recorded by Rondolet before its destruction in the disastrous fire of July 1823 (Rondolet, III:116–119). From Rondolet's drawings and explanatory text, we can discern three different truss systems of different dates. For the earliest truss (which incorporated repairs made to the roof in A.D. 816), constructed of pine, the carpenter devised a double-framed system using paired tie beams and a central hanger as a clasp, whose apex is secured by an iron bolt through the joint of the rafters and whose base is fastened together by a threaded wooden dowel to prevent the tie beam from sagging. The compound trusses support a series of purlins and a ridge piece, which in turn support the roof's outer covering. The trusses are spaced at *bay intervals* of about 3½ meters, while the common rafters within these frames are spaced apart at only 20 cm. In comparison to the total span, the scantling of the major timbers in the roof of St. Paul's Outside the Walls appears exceedingly light, especially in contrast to the post and lintel systems of Greek temples. Moreover, as observed by Rondolet, the doubled truss at St. Paul's illustrates well-developed tension joints necessary for the function of the central hung post that supports the midpoint of the long tie beams.

North of the Alps, the colossal single-naved Constantinian basilica at Trier, dated to ca. A.D. 310 provides an exceptional example of late imperial

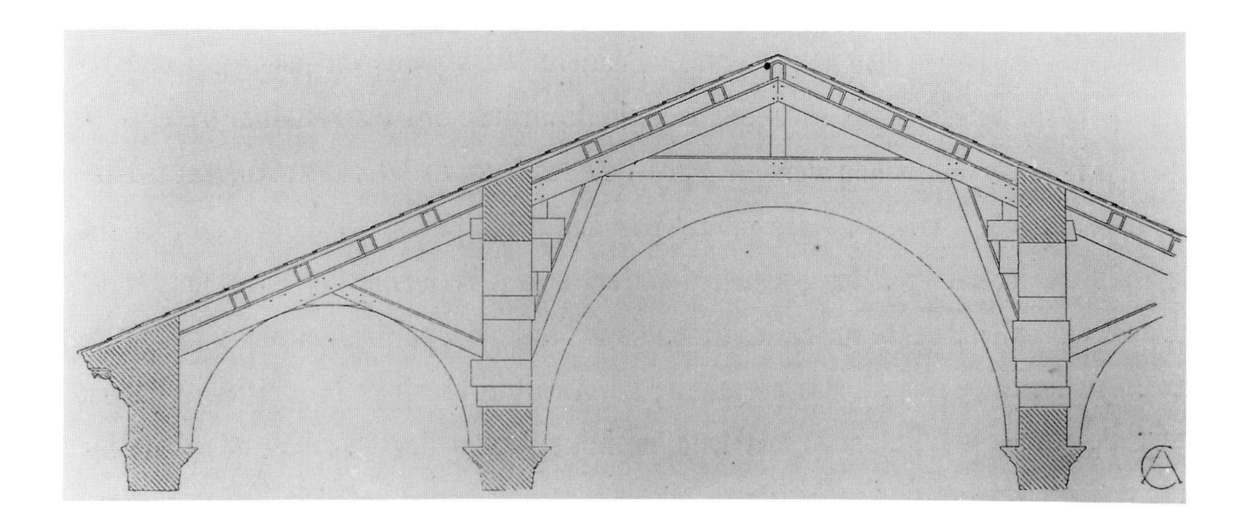

5.15 *Pantheon, Rome: porch roof structure, recorded ca. 1570 (Palladio).*

construction (figure 3.2). The internal span of the main *aula* or *consistorium* is just over 26 meters, which could only have been spanned by a colossal tie-beam truss, and since Trier was a major economic, trading, and shipbuilding center (reached by the Mosel River), great conifers for the tie beams could well have been shipped in. This interpretation is reinforced by the existing support conditions of the masonry. Below the walls, concrete foundations extend 4 meters wide and deep, but the walls above are just 2.7 meters thick (built entirely of fired brick and mortar, with the mortar beds the same thickness as the brick). Moreover, at the level of the window arcading, the upper wall narrows to about 1.7 meters, thus creating a relatively thin-walled support for the enormous-span roof.

Although we are dependent on drawings and descriptions of Christian basilicas for our knowledge of earlier, imperial roof structures, it is possible that the present roof of the porch of the Roman Pantheon, rebuilt in 1632 after Urban VIII removed the "bronze ceiling," may replicate much of the original roof structure, allegedly constructed entirely of bronze (de Fine Licht, 48). The porch of the Pantheon is composed of two major elements: (1) the Greek temple facade with a stone entablature, and (2) a three-aisled hall defined by four rows of two columns set directly behind the outer columns of the facade and the third column from the end, respectively. The columns that divide the porch into aisles all carry axial architraves and a masonry superstructure critical to the support of the roof and the former barrel-vaulted ceiling. The pitch of the roof gable is 25 degrees; the aisle roofs and central span are treated separately, except for the rafters, whose original configuration is impossible to reconstruct. A drawing by Palladio (figure 5.15) shows a heavy-sectioned, continuous rafter carrying close-set common purlins without common rafters. The central main span of the porch (13 meters internally) had a hung ceiling of semicircular curvature. The collar beam spanned only the midportion of the gable, or about 6 meters, which is arguably conceivable to allow fabrication of a member in metal. Likewise, the rafters need not have been continuous because of the masonry support system built upon the architraves both for the central span and the outer columns of the aisles. In sum, the porch roof, whose weight of bronze and nails (of unspecified metal) amounted to more than "450,000 pounds" (Ashby, 202–204) remains of interest chiefly because of the

5.16 *Vitruvius' Basilica at Fanum, interior reconstruction (Morgan).*

concept of triangulated trusses executed in bronze at the time of Hadrian. Whether or not this was actually the case, however, remains in question. It has also been suggested that the entire roof was built of wood covered with bronze (MacDonald, 28), a theory that would explain the almost equal weight of nails mentioned in several accounts as well as the possible need for a stronger framework to support the roof's heavy outer covering, likely of marble.

While Vitruvius' *Ten Books* predates the great expansion of imperial architecture and is generally biased toward Greek prototypes, there are several sections that address the art of roof carpentry and timber as a building material, especially in Book II. He strongly advises, for example, the use of *larch* because of its ability to resist fire (Vitruvius, II. 9. 14–15). Thus it is ironic that Nero in A.D. 56 built a timber-framed amphitheater of exceedingly large larch timbers—a structure described by the Neroean poet Titus Calpurnius Siculus as an immense timber-framed structure "rising to heaven on interlaced beams"—that was nonetheless swept away by the Great Fire of A.D. 64 (Meiggs, 250).

Apart from his description of the basilica at Fanum, roofs and their members as well as various kinds of timber joints are treated in Vitruvius' discussion of machines, such as the war engines described in the latter portions of Book X and also in Book IV, where he mentions tie beams and rafters and, for large spans, crossbeams (i.e., collars) and struts under the roof (Vitruvius, IV. 2. 1). The fact that the rafters extend to the eaves and the mention of tie beams in the framing certainly implies a truss, even though his main focus here appears to be on the origin of triglyphs and metopes in relation to the lintel.

For the history of building technology, the large-scale roof built for the basilica at Fanum is of particular interest because its structure was both designed and commented upon by Vitruvius in Book V. Seemingly a standard Roman structure, the basilica was a two-storied, wide-span building derived from Hellenistic prototypes (figure 5.16). The 120-foot-long roof (all dimensions are given in Roman feet, roughly equal to a modern foot) covers a nave with a span of 60 feet, which when combined with the side aisles gives a total width also of 120 feet. Vitruvius describes what seems to be a base-tied truss resting on columns 50 feet high and 5 feet thick, each having pilasters behind them, 20 feet high and 2½ feet wide, which rise to support the beams that carry the upper flooring of gallery above the aisles; the gallery itself is 18 feet tall. The lower aisle roof is also supported by posts that carry the roof beams supporting the rafters. In the nave, the spaces between the tie beams resting on the columns and pilasters are left open for the clerestory windows. The main frames of the roof, therefore, must have been base-tied trusses. Significantly, Vitruvius' description implies doubled tie beams (*transtra*), which might suggest the same kind of tension connection as seen in Rondolet's version of St. Paul's Outside the Walls (figure 5.7). Members specifically mentioned by Vitruvius are: cross-beams (tie beams or collars), vertical struts, main rafters, longitudinal purlins, common rafters, and *architraves,* which are in essence longitudinal, composite *plates* above the columns composed of "three, two-foot-high timbers fastened together," which project from the side columns of the temple (Vitruvius, V. 1. 4–10).

Reconstructions of the basilica at Fanum differ considerably, but it is likely that Vitruvius

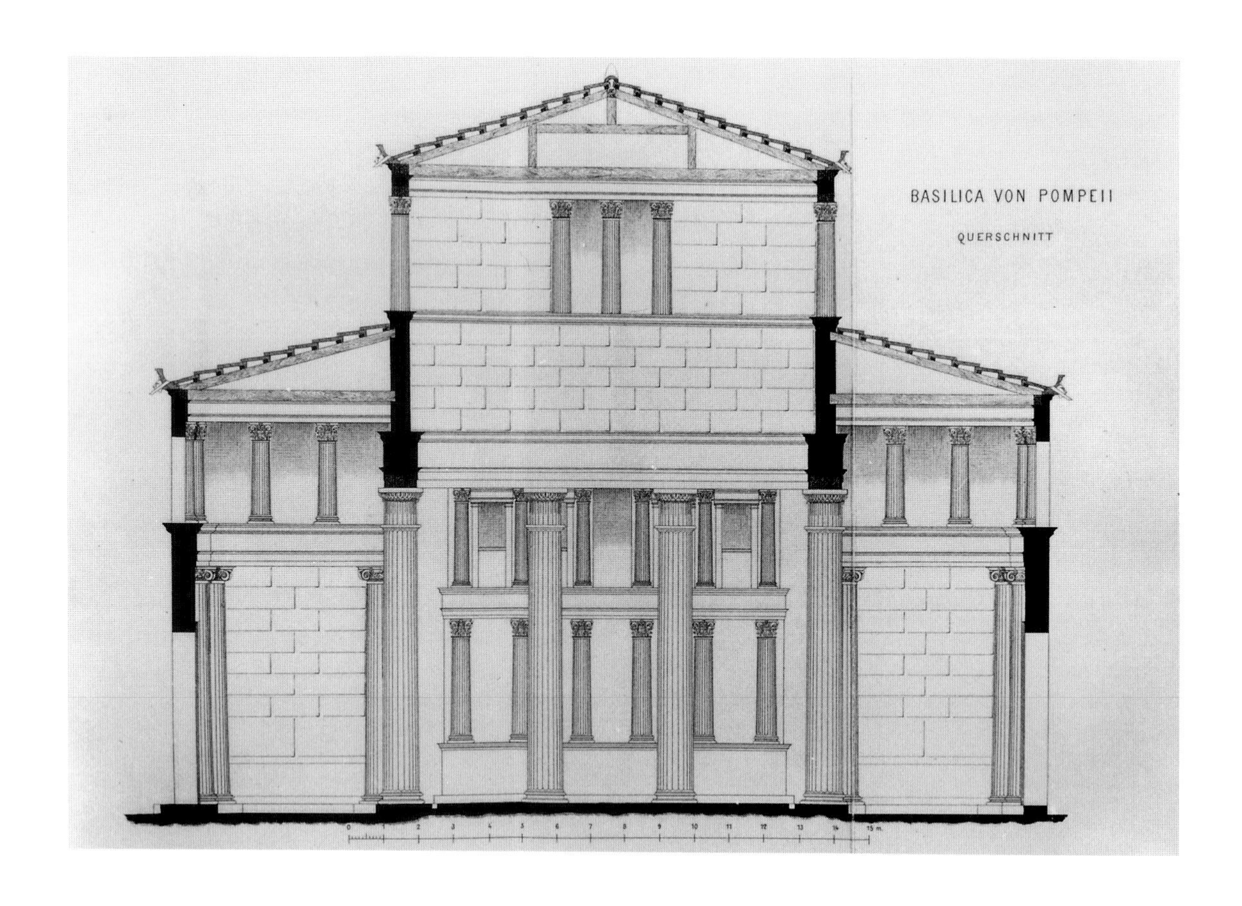

5.17 *Basilica at Pompeii: cross-section, reconstruction (Lange).*

employed a traditional roofing system using principal triangulated tie-beam trusses and purlins. From his description, it appears that column extensions directly support the main-span tie beams, described as *transtra cum capreolis,* or cross beams with *inclined* props or supports (Vitruvius, V. 1. 9). But it is not certain that this is the ubiquitous, hung king-post and double tie-beam construction seen later in Christian basilicas, nor that the *capreolis* are necessarily the diagonal struts that so frequently accompany a hung king post, as suggested by the earliest extant roof of the Monastery of St. Catherine at Mt. Sinai. Vitruvius' text also mentions a simple ridge-pole for a moderate span; and tie beams occur regularly as main roof members (Vitruvius, IV. 2. 1).

Lange's reconstruction of the basilica at Pompeii shows a tie-beam roof that carries two short, vertical struts connected by a crossbeam (i.e., a collar) about midway in the roof gable (figure 5.17), This appears to fit features of Vitruvius' description as well as a roof type that owes its inspiration ultimately to Greek post and lintel construction. Nonetheless, the eminent scholar Walther Sackur contends that Vitruvius' main roof at Fanum contained a hung king post (Sackur, 124–143; Vitruvius, Como ed., folio 54). The truss suggested by Sackur is structurally plausible, and it could very well have been a precursor of similar structures appearing in later Roman basilicas. Indeed, the hung king-post and tie-beam system became a common feature of medieval Italian roofs, and afterwards in northern European roofs; but whether or not it can be attributed to Vitruvius remains speculative.

Roman carpentry technology was surely exported to the provinces of the Empire, as for example, in the variety of joints seen at the recently excavated site of Cannon Street, London (Goodburn), and in the basilica at Trier, whose giant span was probably achieved by means similar to St. Paul's Outside the Walls. The extent to which metal connections were original to joining structural timbers in these lost roofs is, of course, highly speculative, despite Rondolet's observation of the metal bolt at the apex of that building's oldest truss (repaired in A.D. 816 and surviving until 1823; see above) and the extent and general quality of Roman iron production, including the widespread use of large nails.

The legacy of Roman carpentry techniques and the use of the triangulated truss with either wooden joints or metal connections was widely disseminated, from Spain to the British Isles. How much of this technology survived in public buildings or in monastic institutions, or was "rediscovered" in the centuries after the collapse of central power in Rome and the establishment of the various barbarian successor states in the West, remains an unanswered question—especially since the few large roofs known prior to ca. 1100 north of the Alps reveal a different structural approach from the principal truss-and-purlin roofs of the late Empire. Nonetheless, in the Byzantine East, the fortunate survival of the sixth-century roof above the eighteenth-century ceiling at the Monastery of St. Catherine at Mt. Sinai tends to suggest that religious institutions, which conserved classical culture and manuscripts, may also have preserved some of the principles of Roman carpentry, including the invention of the hung king-post truss that appears ubiquitously in France and the Rhineland in monumental roofs of the High Middle Ages.

Byzantine

As indicated in earlier chapters, architects of the eastern empire continued to exploit Roman imperial

5.18 *St. Catherine at Mount Sinai, sixth century nave: roof truss.*

dome construction for which timber centering played an important role. Likewise, basilican churches continued to be constructed in the monasteries of the eastern provinces, as for example, the pilgrimage sanctuary of St. Catherine at Mount Sinai. St. Catherine's monastery, commemorating the alleged site of the miracle of the burning bush and home of desert hermits, was founded and fortified under Justinian between 548 and 565, and survives to this day. The nave roof, the oldest complete timber roof extant, is dated by inscriptions on the tie beams that have also been radiocarbon dated to the mid-sixth century. The carvings and inscriptions on the beams, which were covered by eighteenth-century paneling, give the name of the builder as Stephanos of Aila (Forsyth 1968, 4–9). The roof was thus originally a visible open-timbered structure similar to earlier Roman basilicas in which open trusses and decorative beams rested upon a clerestory wall now illuminated by a series of rectangular windows. Unlike the basilicas discussed earlier, St. Catherine's is not particularly large, no doubt owing to the remoteness of the site and the fact that all of the building materials were imported from distance sources.

The church comprises a typical three-aisled basilica with a nave span, measured in plan from column centers, of 6½ meters. The roof consists of thirteen principal trusses that carry longitudinal purlins (figure 5.18), battens, and roofing material, which was observed in 1384 to be of lead, but was probably originally made of tiles in the Roman manner (Forsyth, 1980, 188). Pairs of rafters (rafter couples) are joined to a central hanging post at the apex and tenoned into the tie beam at its outer end. The central hanging post, which terminates well above the tie, carries diagonal struts that support the midsection of the rafters. The form of the apex of this truss is notable in that the top of the rafter joint also carries a ridge piece, and an extra piece of wood affixed by a large wooden dowel is added to strengthen the apex support. These struts (in compression) are tenoned into the lower portion of the central hanger (which is in tension) and are notched and tenoned into additional timbers nailed to the underside of the principal rafters, apparently to avoid cutting a mortise in the main rafter and weakening it.

In the present ceiling, alternate trusses have metal straps connecting the central hanger to the tie beam, thus recreating the system of tension support at the midpoint of the tie beam observed earlier at Old St. Peter's and St. Paul's Outside the Walls. It is not known how old these straps are, and if they are original to the sixth century. Because of the relatively short span and the small likelihood of problems of bending in the tie beams, the ironwork could very well have been placed when a ceiling was added in the eighteenth century. Yet regardless of the date of the ironwork, which often remains problematic in early timber roofs, these trusses represent an archetypal roof form: the hung king-post truss, with common purlins which continued to be exploited until modern times.

Early Medieval

Almost no monumental roofs survive in western Europe until ca. 1100, after which time the technology of large-scale carpentry exhibits considerable development. Nonetheless, the centuries that preceded the mature Romanesque and Gothic achievements experienced significant structural advances and an enormous quantity of construction. Carolingian ecclesiastical reforms were accompanied by the building of more than 1,200 abbeys, many of which were sizable complexes as at St. Martin at Tours and St. Riquier at Centula in France, and Farfa and St. Salvatore in Italy. In fact, most extant pre-twelfth-century roofs of northern France, Flanders, and the Rhineland represent a rebuilding or expansion of Carolingian foundations after the unsettled period of invasions in the tenth century. These continental roofs are remarkably similar to one another, but they differ significantly from the typical Roman imperial roofs with principal trusses and common purlins. Yet the Frankish architectural and liturgical revival looked to Christian Rome for its inspiration, particularly to the great aisled and trussed-roofed basilicas represented by St. Peter's. The architectural legacy of the Roman timber-trussed basilica, therefore, could well have provided the tradition of roof carpentry for some of the largest buildings of the Middle Ages that date from this period; for example, the churches of Hersfeld and Fulda whose gigantic scale, with internal nave spans of over 16 meters, likely pushed late-Carolingian building technology to its limit (Braunfels, 30–36). Timber-roofed halls like the immense single-nave basilica at Trier or the Roman basilica at Metz, with an internal span of 18 meters, provided colossal *exempla* for the Germanic imperial churches of the eleventh-century (Oswald, 157). But the earliest extant roofs in northern Europe do not have main trusses or purlins and instead reflect a general type of common-rafter truss with tie beams for every pair of rafters, generally spaced a meter apart or less as, for example, in the early roofs of the nave of St. Vincent at Soignies in Hainault, Belgium, and St. Gertrude at Nivelles, in the ancient Duchy of Brabant.

Although the early carpentry of the church of St. Gertrude at Nivelles, constructed in 1046 with a nave span of just over 10 meters, no longer exists, there is substantial evidence from the imprint on the east choir gable from which the eleventh-century roof has been reconstructed. The earliest pitch was 30 degrees and the construction very simple, similar to the truss illustrated in figure 5.9, without a collar or a hung king post, but with four slightly canted struts between the rafters and the tie beam (rather than the two shown in figure 5.9). Since the roof was originally enclosed with oak paneling (Lemaire, 250), the tie beams also supported axial timbers for the ceiling—a common system that has been reconstructed in the late-eleventh-century painted ceiling of the Abbey of St. Michael at Hildesheim (Bohland). At Nivelles and related roofs, the tie beams rest directly on the clerestory walls, without wall plates. Significantly, despite its relatively large scale, the roof remains totally devoid of longitudinal bracing, as is characteristic of the earliest recorded roofs well into the mature Romanesque.

The late-eleventh-century roof of the nave of St. Vincent at Soignies, about 11 meters in span, exhibits structural development over that of St. Gertrude with the addition of a transverse collar beam, an increase in roof pitch to 45 degrees, and longitudinal wall plates that secure the rafter feet and more uniformly distribute rafter loadings to the top of masonry walls (figure 5.10)—features that become standard in Romanesque roofs with uniform scantling. Also characteristically, the tie-beam and rafter trusses are closely spaced, just under a meter from center to center. The frame is stiffened in the transverse plane by the vertical struts with pegged, notched lap joints into the rafters and by the half-lapping of the collar and innermost posts.

The nave roofs of Soignies and Nivelles are typical examples of early Romanesque roofing systems in large churches from about the end of the first millennium on, and provide the immediate structural and typological background for most large-scale English and continental roofs until ca. 1270 (Fletcher and Spokes, 155–158). They comprise part of a large category of uniform-scantling roofs without bays that place evenly distributed loads along the wall head, as opposed to the *double framing* seen north of the Alps, initially in Norway where stave church roofs are framed by heavy principal rafters and longitudinal purlins forming bays.

The Norwegian heavy principal-rafter roof with long diagonals called *scissor braces* and longitudinal purlins and ridge pieces constitutes a special category of medieval roofs that seems to have emerged initially in the timber halls and churches of Scandinavia constructed with earth-fast posts. The oldest surviving timber-framed structure of this general type is the Saxon church of Greensted, Essex, incorporating *palisade* wall construction traditionally dated to the ninth century (Brunskill, 189). How the building was roofed is unknown, but originally the large, vertical halved timbers called *staves* or *masts* were inserted into post holes in the ground—a form of timberwork typical of much early domestic and ecclesiastical building prior to the evolution of the fully developed framed stave churches of Norway.

Of particular interest for early roof construction, in Britain and northern Europe, however, is the recently demolished Saxon roof of Odda's Chapel at Deerhurst. As an example of late Saxon roof carpentry, the chapel, dated to ca. 1056, shows parallels with the Continent and Denmark. The roof, pitched at about 45 degrees, was composed of trusses with tie beams, king posts, common rafters, and canted braces connecting the rafters and the ties that spanned the full width of the wall, a length of only 5½ meters (Currie, 59). Odda's chapel, however, lacks the ridge piece and principal rafter system of the scissor-braced trusses normally associated with later Norwegian stave churches, whose character is well documented. These have double-framed roofs with principal trusses that support boarding rather than common rafters, *scissor braces* (diagonally crossed timbers) that spring from the feet of the principal rafters (figure 5.19), and joining techniques that differ mainly from the normal family of English and continental rafter roofs of the eleventh and twelfth centuries in that lap joints are unnotched.

With their sophisticated double framing, bay construction and attention to detail, it is generally held that the basilican **stave churches** of Norway represent the highest level of roofing technology in the North at the end of the eleventh century, even though they do not compare in scale with much larger "long houses" recently excavated, for example at the Viking village of Saedding. Nor do they even approach the span of the great abbeys at Fulda, Hildesheim, or Soignies. Nonetheless, Norwegian stave church roofs, which evolved to their mature form by 1200, display the important technical application of principal rafters in doubled-frame post and truss construction with ridge pieces and purlins found nowhere in England or continental Europe from ca. 1050–1300 (Smith 1977). More critical for the history of medieval roofing technology are the innovations that produced both longitudinal and vertical integration of the roof structure by means of purlins and effective bracing techniques (for example, the scissor braces and short curved braces illustrated in figure 5.19), which when applied at the appropriate intersections of main upright and cross-beams helped prevent both lengthwise raking and

5.19 *Borgund stave church, Sogn, ca. 1250: roof framing.*

lateral distortion. In short, the Norwegian use of principal trusses, effective yet simple curved bracing, and longitudinal purlins prefigures a considerably later development of principal-rafter Gothic roofs in England of considerably larger scale. Yet it is to the tradition of simple common-rafter roofs of uniform scantling that we must return to assess the technical advancement of mature Romanesque roofs.

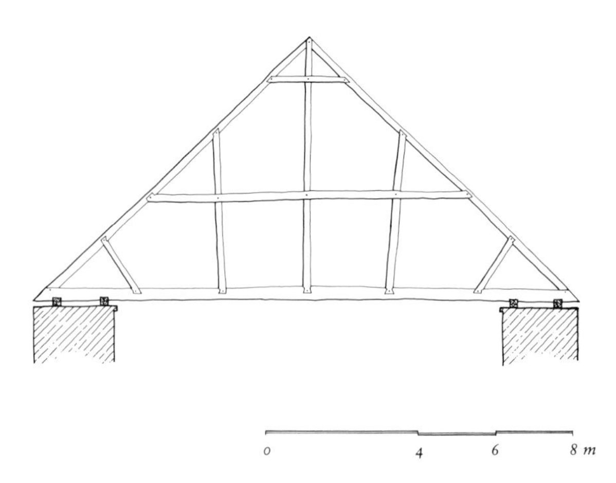

5.20 *Ely Cathedral: Romanesque nave roof, reconstruction (after Simpson).*

Romanesque

Two critical factors contributed to the technological evolution of roofing systems in the late eleventh and twelfth centuries. First, the increase in large-scale building reflected a period of general prosperity, economic and political stability, and religious zeal. All categories of building enterprise, military and civil as well as ecclesiastical, experienced accelerated levels of activity in northern Europe. Second, the greatly increased use of stone vaulting, often to replace earlier timber roofs and ceilings, resulted in a need to accommodate masonry vaults that intruded into the region between the clerestory walls that was normally occupied by closely spaced tie beams. As observed earlier, particularly as churches of the twelfth century became taller and with the concomitant necessity for supporting roofs above vaults on relatively thin walls, carpenters were challenged to cope with higher wind loads acting on the roofs, while at the same time they had to find means to decrease the number of close-set tie beams, both to avoid the vault crowns and to economize on the expenditure of long, straight timbers.

The earliest Romanesque roofs in northern Europe, like their Ottonian predecessors, are not above vaults but rather are open-timbered or intended to support a paneled ceiling. As with their antecedents, these roofs today are generally fragmentary survivals of various repairs and rebuilding, witnessed by reused timbers, empty mortises, and imprints of earlier timbers on masonry walls. Among the important Anglo-Norman ecclesiastical roofs of this period whose archaeological remains offer possible reconstruction are the nave of Ely Cathedral, and to a lesser extent the nave of the abbey church at Jumièges. Both roofs spanned 10 meters and were of about 40 degree pitch, and both were the pervasive common-rafter type with tie beams in every frame. Dendrochronological information from reused timbers in the present nave of ca. 1250 scars and on the east face of the west tower at Ely Cathedral (the only Norman nave for which this information exists) provided the basis for a reconstruction of the design of the original Romanesque roof (Simpson).

The nave roof of Ely was originally supported by about ninety triangulated collar-rafter tie-beam trusses spaced at intervals of just under a meter (figure 5.20). As at Odda's Chapel (above), a central post rises from the tie beams (of oak, about 30 × 30 cm. in section) to the roof's apex without a ridge

5.21 *Abbey Church, Jumièges: upper, west face of the crossing tower.*

piece. Unlike earlier Ottonian roofs (e.g., Nivelles, and the nave of Soignies), the Ely roof may have had double wall plates and triangulated feet. As with all the roofs of ca. 1050–1200, the joints are of the simple lap type, with or without notching or dovetail profiles. Dendrochronology suggests a date of ca. 1120 for the original nave, thus classifying it as one of the earliest known cathedral roofs in England (Simpson). Moreover, the reconstruction of Ely provides important clues to the general nature of the earlier Romanesque roofs in England as well as in Normandy, including the roof of the nave of Jumièges.

As noted earlier, the unusually tall Romanesque nave of the abbey church at Jumièges marks a significant stage in the evolution of monumental ecclesiastical architecture prior to the arrival of Gothic structure in northern France. Although attention has naturally focused on the imposing extant masonry, considerable archaeological evidence in the stone informs us about the several roofs that once surmounted this magnificent edifice. From the west face of the crossing tower (figure 5.21), it is evident that the church has had three roofs of different pitch: close to 20, 40, and 55 degrees respectively. Of the three, the lowest pitch likely dates from the seventeenth century and relates to the inserted timber vault of 1692. The highest roof scar is also not original, since its steep pitch places it stylistically in the Gothic period. More telling, the "Gothic" roof cuts into the lower portions of the Romanesque tower windows, leaving the middle roof as the only choice for the Romanesque nave.

Additional archaeological evidence indicates that the earliest roof above the nave lacked a ridge piece. Evenly-spaced tie-beam sockets on the inner faces of both western towers and intermittent corbels suggest housings for a series of rectangular tie beams, about 26 × 31 cm in section, and spaced close to a meter apart. Since all the sockets are the same size and located at a level corresponding to the base of the Romanesque gable wall, one can infer that the original eleventh-century roof comprised a series of uniform tie-beam trusses, possibly with vertical bracing to the rafters. The base of the roof gable on the east face of the crossing tower (figure 5.21) contains slots intended to house longitudinal timbers running at least from the last transverse tie beam crossing the nave, suggesting that axial timbers were intended to carry a ceiling, in the tradition of Ottonian churches. If so, the original roof of the nave of Jumièges differed from the Romanesque nave of Ely, which displays no evidence for the existence of a ceiling.

In the latter part of the twelfth century, Romanesque roofs above vaults, while generally similar to those already discussed, exhibit an evolution toward bay-type construction accompanied by greater experimentation in bracing, a gradual decrease in the number of tie beams used at the base of the frames, and the associated development of the *triangulated rafter foot* (figure 5.22). In the roof of the vaulted choir of St. Vincent at Soignies, for example, traditionally dated to ca. 1170 (and like so many others, a rebuilding of the tenth-century structure; Brigode, 150), the tie-beam trusses occur only at every third frame. The trusses, with a collar and five vertical struts, including a central post to the apex without a ridge piece, are thus the *primary trusses within a common-rafter system*. By comparison, the intermediate frames are less elaborate, but they incorporate an enlarged triangulated rafter foot

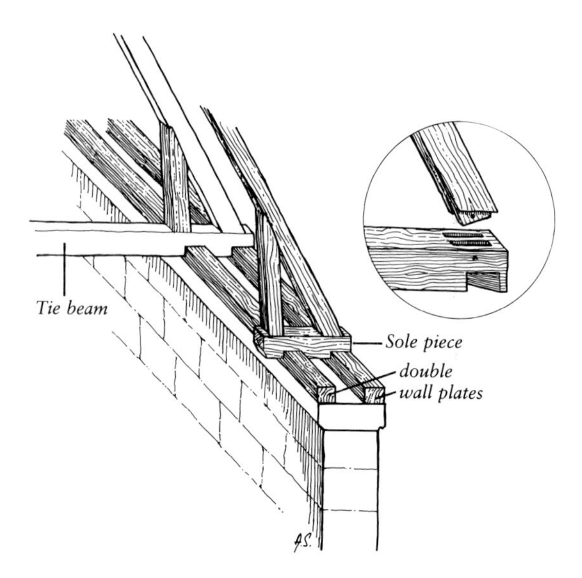

5.22 *Triangulated rafter foot at the base of an early Gothic roof frame with shouldered mortice and tenon joint (after Viollet-le-Duc).*

bearing at its inner end on a longitudinal plate, sometimes termed a *flying plate* (figure 5.23). Perhaps most significant, this additional longitudinal plate in combination with the extended rafter foot reveals a level of technical understanding of structure that anticipates the later development of longitudinal bracing with purlins used in Norwegian churches and the roofs of Gothic cathedrals.

The late Norman roof of the nave of the cathedral of Lisieux (figure 5.24), whose original construction of ca. 1180 was rebuilt according to the original design after a fire in 1226, offers an early example of the application of *scissor-bracing* and doubling of the lower rafters in an attempt to achieve rigidity without the need for tie beams at the base of every frame. In both nave and transepts (spans 8 and 9 m respectively), the scissor braces and extra, lower rafters (*secondary rafters*) are joined at each end by pegged and notched lap joints capable of accepting tension. This has allowed the tie beams to be widely spaced (and uneven), about every seventh

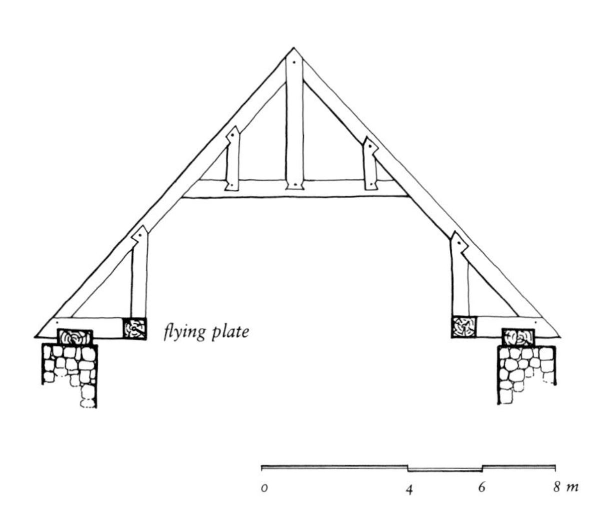

5.23 *St. Vincent of Soignies: intermediate framing of the choir, ca. 1170.*

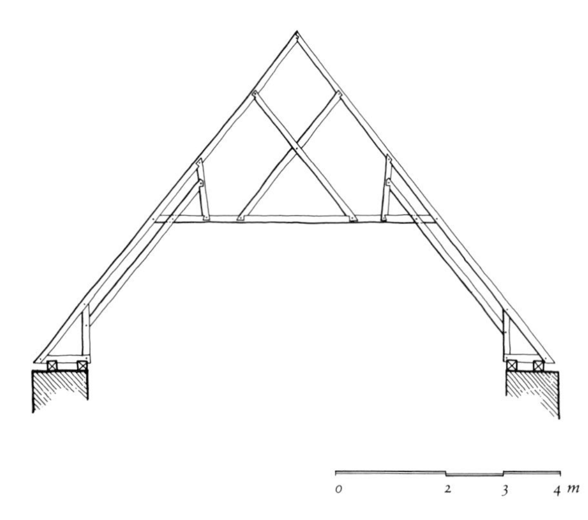

5.24 *Lisieux Cathedral: transept framing of ca. 1180 with scissor brace, rebuilt after 1226 (drawing after CRMH).*

pair of scissor-braced rafter couples, which in turn are but half a meter apart in the nave. The ties are trenched over two wall plates, and the stout section of these intermittent ties (total nine) serves to stabilize the wall plates as well as the upper wall, functioning as anchor beams for the masonry (Ostendorf, 14).

The roofs of the transepts of Lisieux and the choir of Soignies share a common technology. Both employ almost exclusively heartwood timber, light and uniform scantling, and simple lap and notched-lap joints. They also exemplify four important developments that frequently occur within the family of common rafter roofs: (1) the gradual elimination of tie beams at the base of every frame; (2) the use of supplementary bracing to stiffen the roof frame; (3) the development of a triangular rafter foot; and (4) supplemental rafters, which can greatly increase the rigidity of the common rafter system above the rafter foot. It is also evident that Romanesque carpenters of the later twelfth century tended to evolve a type of "bay framing" within a uniform system of base-tied trusses, even though effective longitudinal bracing was still limited to the region of the wall plates.

Gothic

From the mid-thirteenth through the fourteenth century, a wide range of roofing systems evolved to cope with both wide spans (over 14 meters) and the structural requisites of tall Gothic cathedrals and their spires. During this period, too, roofs became generally steeper in pitch, on average, about 60 degrees, a transformation that parallels the adoption of the Gothic pointed, ribbed vault (chapter 4).

Mainly in the first half of the thirteenth century, as demonstrated in many instances of both roofs over vaults and open-timber roofs, the single most important technological change was the **longitudinal integration** of formerly independent roof trusses achieved variously by strengthening the primary rafters and adding longitudinal purlins, a ridge piece at the apex, and diagonal members providing lateral reinforcement, in the plane of the roof called *wind bracing*. This bracing was initially realized by straight timbers, set diagonally, and later by curved braces, which in open roofs like Westminster Hall provided decorative enrichment as well. Gothic roofs above vaults generally became light, efficient systems designed to work with the new support conditions of thin-walled masonry. In open-timber roofs, especially in England, principal rafters evolved in a different structural context in which traditional aisled construction, crucks, and hybrid combinations merge to produce a rich variety of roof forms, some of which recall the doubled-framed construction of the earlier Scandinavian tradition.

As in vaulted Romanesque churches, Gothic roofs above vaults involve a particular set of structural criteria because of the intrusion of the extrados of the vaults into the space normally occupied by tie beams. It was both in the interest of economy and to meet the need to accommodate masonry forms that carpenters were encouraged to seek means to reduce the number of tie beams across the base of a roof. To do this, stronger, base-tied trusses were developed with additional lateral and longitudinal bracing such as the secondary rafters initially seen at the Cathedral of Lisieux. In some cases, too, tie beams were entirely eliminated as part of that era's experimentation in roofing systems above vaults.

The essence of these longitudinally integrated roofing systems can perhaps be traced most clearly in the nave of Notre Dame in Paris whose roof of ca. 1200, possibly designed as early as ca. 1180, was likely erected a few decades prior to the general remodeling of ca. 1225–1235 (see chapter 3). Though repaired a number of times, there is no conclusive evidence that either the basic design or most of the major members of the nave roof were significantly altered (Courtenay 1989, 57). Hence we can regard the Notre Dame nave as offering a seminal northern French example of solutions to the problem of roof stability in a tall building.

The framing embodies two kinds of trusses: main frames corresponding to every fifth rafter couple with tie beams (figure 5.25), and secondary rafter couples without ties, spaced at 80 cm, with a traditional rafter foot (as seen in figure 5.26). Because the roof framing was erected *before* the construction of the vaults below, it provided work platforms for lifting the materials needed for vault construction. The primary elements allowing the frame to be used in this manner are the hangers, which clasp the large oak heartwood tie beams with efficient tension joints (figures 5.25 and 5.27) that are replicated also at higher points of attachment. Longitudinal bracing, below the rafter foot and within the plane of the roof, can also be observed in figures 5.26 and 5.27.

Notre Dame in Paris provides still another example of the general tendency for carpenters to develop main frames or bay division of supports within a common rafter system. In short, all the main rafter couples of Paris remain of uniform scantling, but the important particulars of their design and function during vault construction marks a clear differentiation between the main tie-beam trusses with hangers and secondary rafters and the ordinary rafter couples.

The unusually steeply pitched (almost 70 degrees) roof of Reims Cathedral, whose carpentry, rebuilt after the fire of 1481, was recorded by Viollet-le-Duc (figure 5.28) and others before its total de-

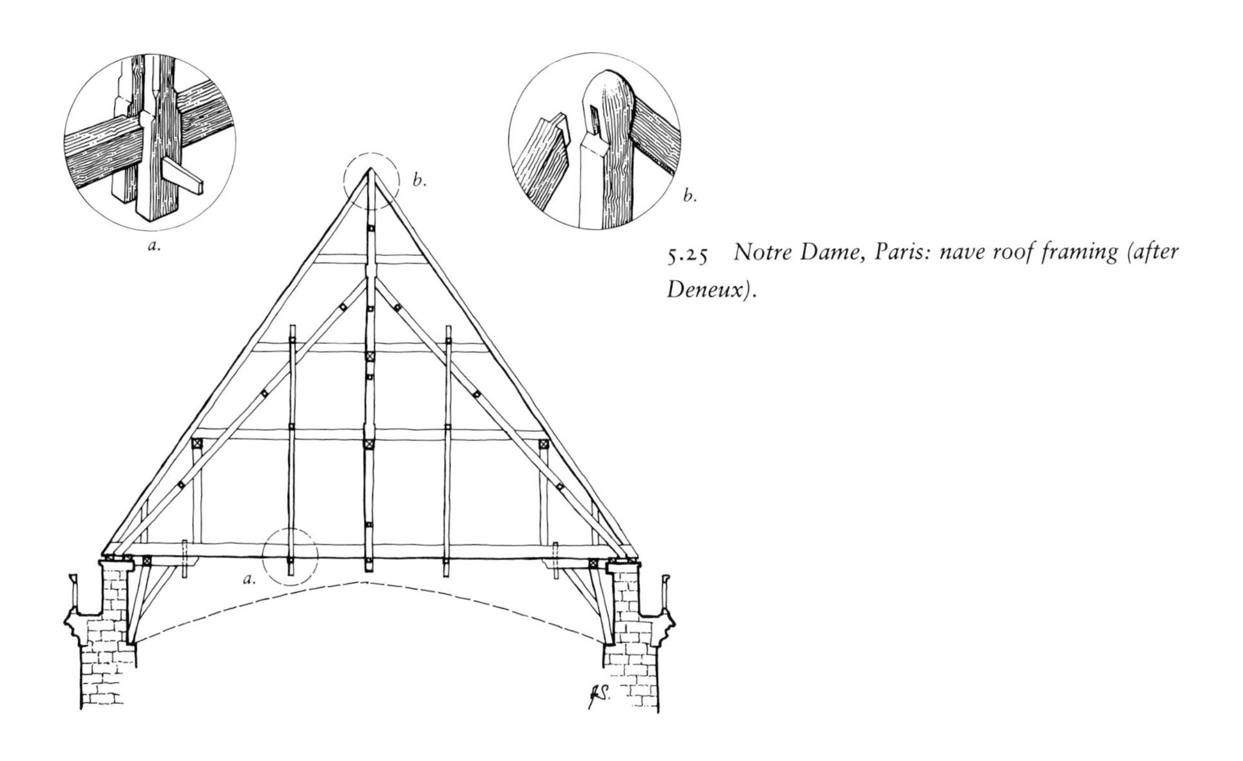

5.25 Notre Dame, Paris: nave roof framing (after Deneux).

5.26 Notre Dame, Paris: nave roof rafter foot with wall post.

5.27 *Notre Dame, Paris, tension joints in nave, lateral roof hanger, north side.*

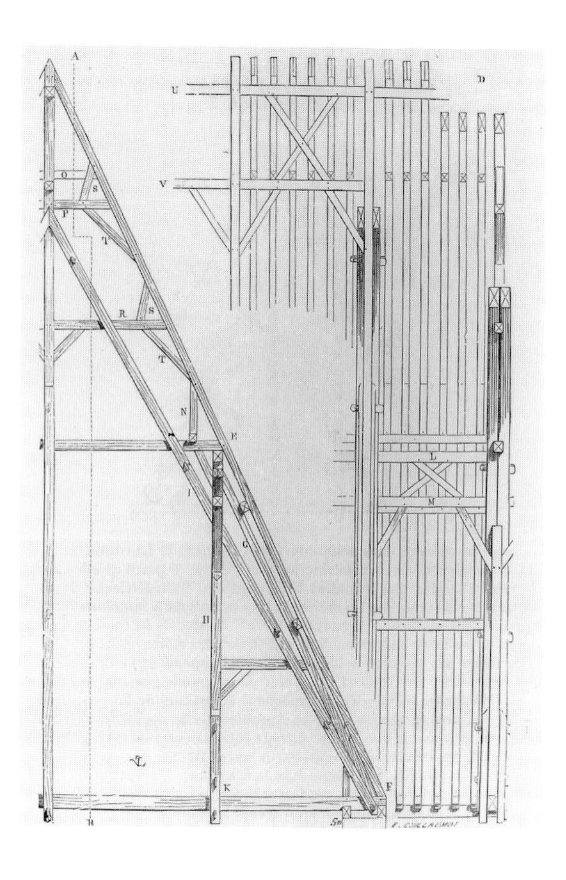

5.28 *Reims Cathedral: nave roof framing (Viollet-le-Duc).*

struction in the First World War, spans nearly 15 meters to the outer parapet wall. Its construction illustrates similarities with Notre Dame in Paris in that it has, like most High Gothic cathedral roofs, main tie-beam trusses within the traditional system of uniform scantling. The main frames, however, occur only at every seventh rafter couple. Like Paris, and other thirteenth-century roofs of northern France, the main truss has a central hung post that carries longitudinal members. The tension joints, made with parallel timbers and wooden dowels, are also similar to Paris. The roof at Reims has additional longitudinal plates, set square and supported by a vertical post, as at Paris, but at Reims the fifteenth-century carpenter has also used inclined purlins set between the outer and secondary rafters. Here too, the transverse plane of the roof is strengthened by a long, slanted brace that extends from the base of the tie beam to just beneath the upper, horizontal collar, as at Lisieux. This single member crosses four other horizontal and vertical roof members and thus constitutes what is known as a *passing brace.* The light, elegant framing of Reims illustrates the efficiency achieved in this era with relatively small-sectioned timbers. The cross-sections of most major members except for the tie beams, that is the rafters, diagonal braces, collars, etc., are square with an average dimension of 22 cm on a side; the tie-

5.29 *Salisbury Cathedral: Nave roof framing.*

beam's scantling is almost twice that dimension. Thus the aspect ratio of the scantling to the longest timbers of about 15 meters is close to: 1:75—in extraordinary contrast to the timber construction of ancient Greece, but not atypical for ecclesiastical roofs in northern France and England, where a similar system of light-sectioned timbers were used in thirteenth-century roofs above cathedral vaults (for example, at Salisbury, figure 5.29).

The aesthetic exploitation of open-timber roofs is a salient feature of English Late Gothic architecture that parallels, and even rivals, the textural effects achieved in vaulting of the same period (chapter 4). Within the general progression of medieval

5.30 *Westminster Hall: interior.*

roofs, no single work represents more clearly the integration of technical virtuosity with iconographic and aesthetic goals than the late fourteenth-century roof of Westminster Hall in London. But even more important in the context of this chapter, the Westminster Hall carpentry displays one of the finest examples of longitudinal structural integration that, by means of augmented purlins, an apex ridge piece, and arched wind bracing, constitutes the structural culmination of the late medieval *principal-rafter roof* (see figure 5.5). In this system, the common rafters derive almost their entire support from the longitudinal purlins, so that their reaction on the wall head is almost negligible—so slight, in fact, as to allow great openings for the illustrated seventeenth-century dormer windows in the plane of the roof.

Designed and executed by Richard II's chief carpenter Hugh Herland at the close of the fourteenth century, Westminster Hall is by far the largest of its type. Built in but three years, from the felling of the timber to the erection of the framing, the roof spans the vast eleventh-century Norman Hall that was originally constructed with internal arcades and gallery as part of the Anglo-Norman palace (figure 5.30). Once the Norman timber arcades were removed, the hall had an unprecedented span of nearly 21 meters, exceeding by more than 6 meters the widest un-aisled hall in medieval England (the great hall at Kenilworth Castle constructed between 1389–1393; Courtenay 1984, 302).

The roof consists of thirteen principal frames (an estimated 600 tons of oak) that divide the hall into twelve bays at 5½-meter intervals. While the overall scale seems vast, it is the uncommon span, achieved with such elegance, that makes the hall unique and has evoked considerable commentary on the design and structure. Even so, uncertainty about just how its complex structural elements function had remained unresolved until its recent modeling, using an instrumented small-scale timber model, and subsequent archaeological evidence.

The modeling revealed that the principal load-bearing frames are supported mainly at the level of the masonry corbel, about halfway down from the top of the wall (figure 5.31), by the combined action of the heavy hammer posts and the great arch which, although composed of pinned-together elements, acts as a single element to convey much of the vertical dead load and the horizontal thrust received from the hammer posts and hammer beams respectively (Courtenay and Mark). The hammer beams were found to carry appreciable *tension,* in contrast to all the other major elements of the frame acting in compression, and thus restrain the outward

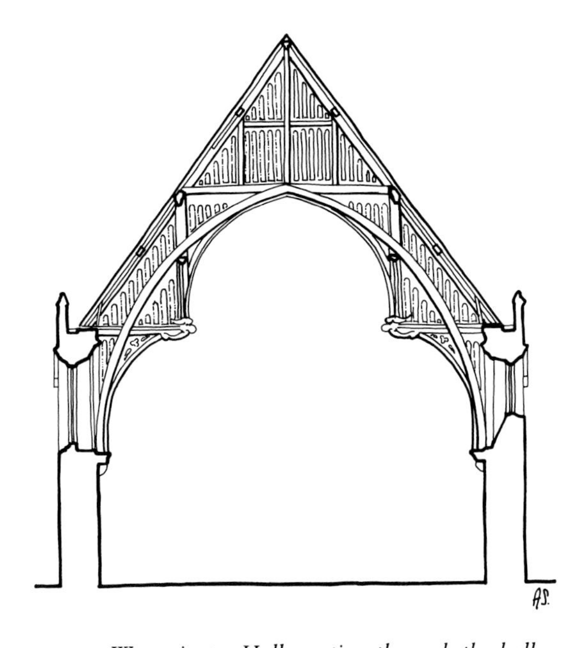

5.31 *Westminster Hall: section through the hall.*

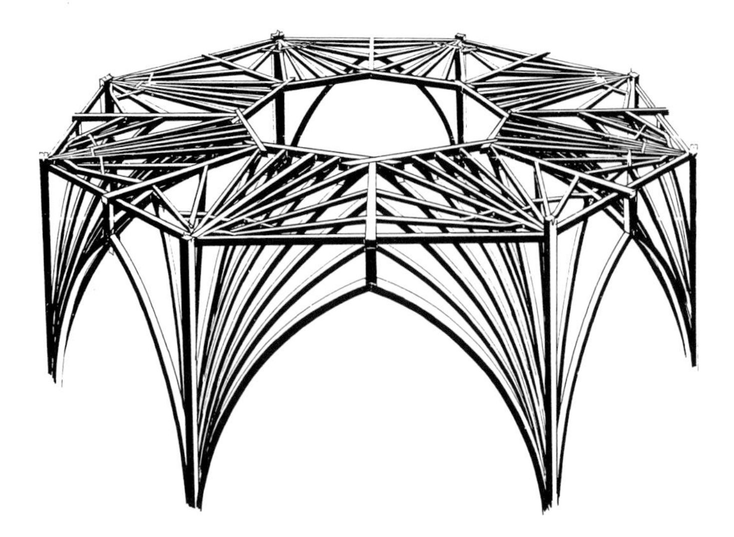

thrust at the base of the rafters. These findings indicate that Herland intended to bring the reactions of this immense roof lower down on the Norman walls, where they are better able to resist horizontal forces—in effect, by reducing the moment "arm" (i.e., reduce the distance *y* in figure 3.6) and decrease the overturning moment which, if not so controlled, could lead to tensile cracking in the supporting walls. This inference is confirmed by Herland's design that created a clear space behind the upper half of the wall post, ensuring that it would not bear against the wall behind it (Courtenay 1990).

While their structures are quite different, the structural and aesthetic integration of masonry and carpentry at Westminster Hall parallels the framing of the complex timber vaulting above the crossing of Ely Cathedral, constructed in 1328–1335. The timberwork supporting the wooden vaults, spanning some 22 meters, depends upon great angle posts set into the corners of the stone octagon, whose grooves are still discernable. The vault and its lower brackets also form a braced timber floor from which the eight posts of the timber lantern rise, supported by enormous raking struts resting on the masonry piers below. The base of the lantern comprises a three-dimensional structure that derives stability

5.32 *Ely Cathedral: timber framing of the tierceron vaults over the crossing (Hewett).*

from ring beams and brackets, as shown in figure 5.32. Like the wall posts of Westminster Hall, the timber framing of the Ely crossing and lantern tower is carefully fitted into the masonry octagon with grooves and sill hooks (Hewett, 160–163), indicating that the timberwork and masonry were designed together to create the centralized, vaulted choir which, for iconographic reasons, looks like a masonry rendering of sacred space.

Aesthetically and symbolically, the tall, slender **spires** of Gothic cathedrals expresses the spiritual aspirations of an age of piety as well as the technological ascent of masons and carpenters. As was generally the case, the spire (or "flèche") erected over the crossing juncture of the nave and choir of Notre Dame in Paris also served as a bell tower, containing six small bells. It survived, albeit in damaged condition, from its construction in the early thirteenth century until its mutilation during the French Revolution in 1792–1793 and subsequent removal by an early-nineteenth-century cathedral architect (the present spire by Viollet-le-Duc, figure 5.33, dates from 1860). Nonetheless, considering the

5.33 *Notre Dame, Paris: model of the nineteenth-century spire base by Viollet-le-Duc.*

usual fate of medieval spires—for example, those of Reims and Amiens had long since perished from violent storms, or fire ignited by bolts of lightning, as at Rouen in 1822—it is extraordinary that such a perilous structure could survive for 500 years. Perhaps even more remarkable, this important example of early Gothic timberwork was recorded before its removal as part of the mid-nineteenth-century major restoration of the cathedral.

While precise dates of the different campaigns of construction of the cathedral's superstructure are unavailable, the spire was most likely erected in the opening decades of the thirteenth century. The original flèche may have been the earliest of its genre. It was different from the majority of twelfth- and thirteenth-century crossing towers that were either mainly of stone, as for example, at the cathedrals of Laon and Strasbourg, or those whose construction coupled a timber spire to a visible, stone base, as formerly at Rouen Cathedral, Mont St. Michel, and Amiens; the latter's present spire (figure 5.34), the oldest extant in France (rebuilt between 1529–1532), rose to a height of 110 meters (Durand, 525). And like Paris, the entire construction, except for the short stone base, is timber covered with lead. While the framing of the sixteenth century may differ somewhat from the original construction of ca. 1240, the spire of Amiens nonetheless shows certain affinities with the thirteenth-century work at Notre Dame in its diagonal bracing from the crossing piers and the slender proportions of the octagonal pyramid. The crossing also illustrates the important interaction between masonry and carpentry in the use of a purpose-built masonry base to support a platform of braced tie beams from which the octagonal spire ascends. The base of the Amiens spire rises abruptly from the adjacent roofs without the elegantly inte-

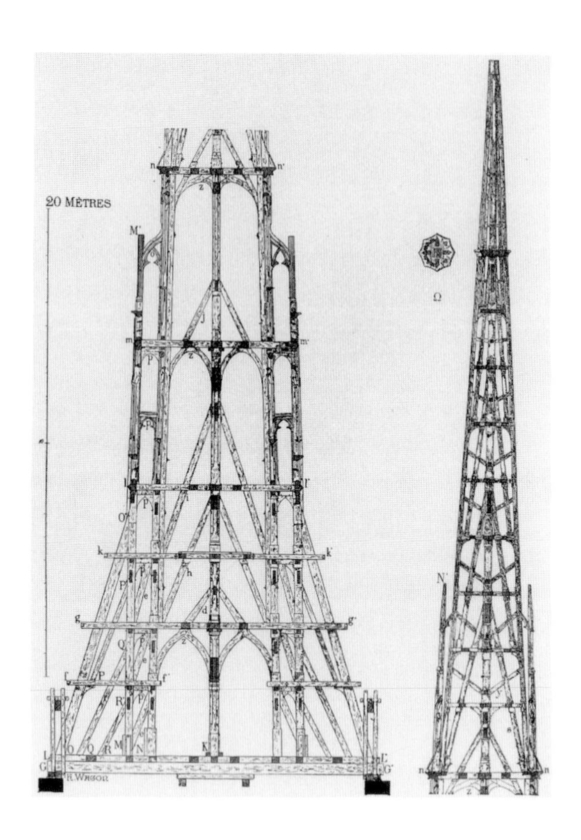

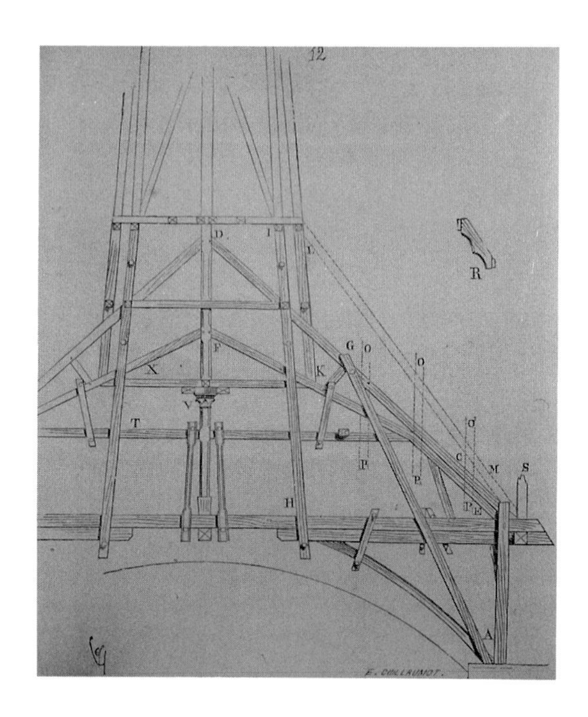

5.34 *Amiens Cathedral: structure of the crossing spire (Viollet-le-Duc).*

5.35 *Notre Dame, Paris: structure of the thirteenth-century spire (Viollet-le-Duc).*

grated visual and structural transition of the Paris prototype.

The flèche of Notre Dame, on the other hand, was a structure conceived at the outset entirely of timber having a base, a multiple-story spire, and pyramid roof, emerging *directly* from the intersecting roofs of the nave and choir roofs. Indeed, the distinctive feature of the original spire was its framed base, sprung from the crossing piers well below the top of the parapet walls. The rigidity of this structure was assured by heavy slanting braces and diagonal ties (figure 5.35), and prominent brackets and hangers with clasp-like tension connections (to join members both vertically and horizontally) that had earlier been developed for such long-span roof structures as the Notre Dame nave. The shape of the spire, a tapered octagonal pyramid, helps also to decrease both mass and lateral wind loading at higher elevations. Given the fact that Notre Dame was a daringly tall, thin building at the end of the twelfth century, and that the spire reached an overall height of approximately 77 meters above the ground, the structure indicates a high level of understanding on the part of the medieval builders of wind effects, which in a less rigid structure might have produced severe rotation and deflections.

Renaissance

The widespread dissemination of classical roof designs (e.g., the Roman-type trusses seen in early Christian basilicas) in publications of the Renaissance, especially by Andrea Palladio in his *I Quattro Libri dell' Architettura,* exhibit the continuity of many of the traditional timber roofing systems south of the Alps. Early triangulated forms that evolved in large-scale imperial roofs with hung tension members supporting tie beams endured well into the modern era, partly from their traditional appeal, but also because of their efficient technology. It was not until late in the seventeenth century that roofs began to reflect the change from the almost universal use of wooden joining to an assembly designed specifically for metal connections using threaded metal bolts and iron strapwork. Yet even into our own time, the traditional forms continue to be employed.

Both from writings and observing contemporaneous structures, we know that a recurring problem concerned the availability of large timbers for roofing long spans. For example, Alberti writes in ca. 1450 that "if the trees are too small to make a complete beam out of a single trunk, join together several into a composite beam . . . [with] their opening faces notched into each other" (Alberti, III. 12. 47–48). Brunelleschi, too, whose scaffolding, hoisting machinery, and timber reinforcement for the great dome of Florence Cathedral (perhaps the most ambitious use of structural carpentry in the Italian Renaissance) was delayed for nearly two years by the difficulty in delivering suitable chestnut beams ordered in 1421 (Saalman, 117).

Philibert de l'Orme (1510–1570), who like Alberti was a writer and theorist (see chapter 6 on the "new professionalism") as well as royal architect to Henry II of France, published a two-volume treatise in 1561, *Nouvelles Inventions pour bien bastir,* entirely devoted to the subject of carpentry—a work that was to become influential especially in the eighteenth and early nineteenth centuries. Specifically "new" in Philibert's proposed structures was the complete abandonment of traditional timber framing with triangulated beam-like members in favor of segmental arches built up from relatively small timber

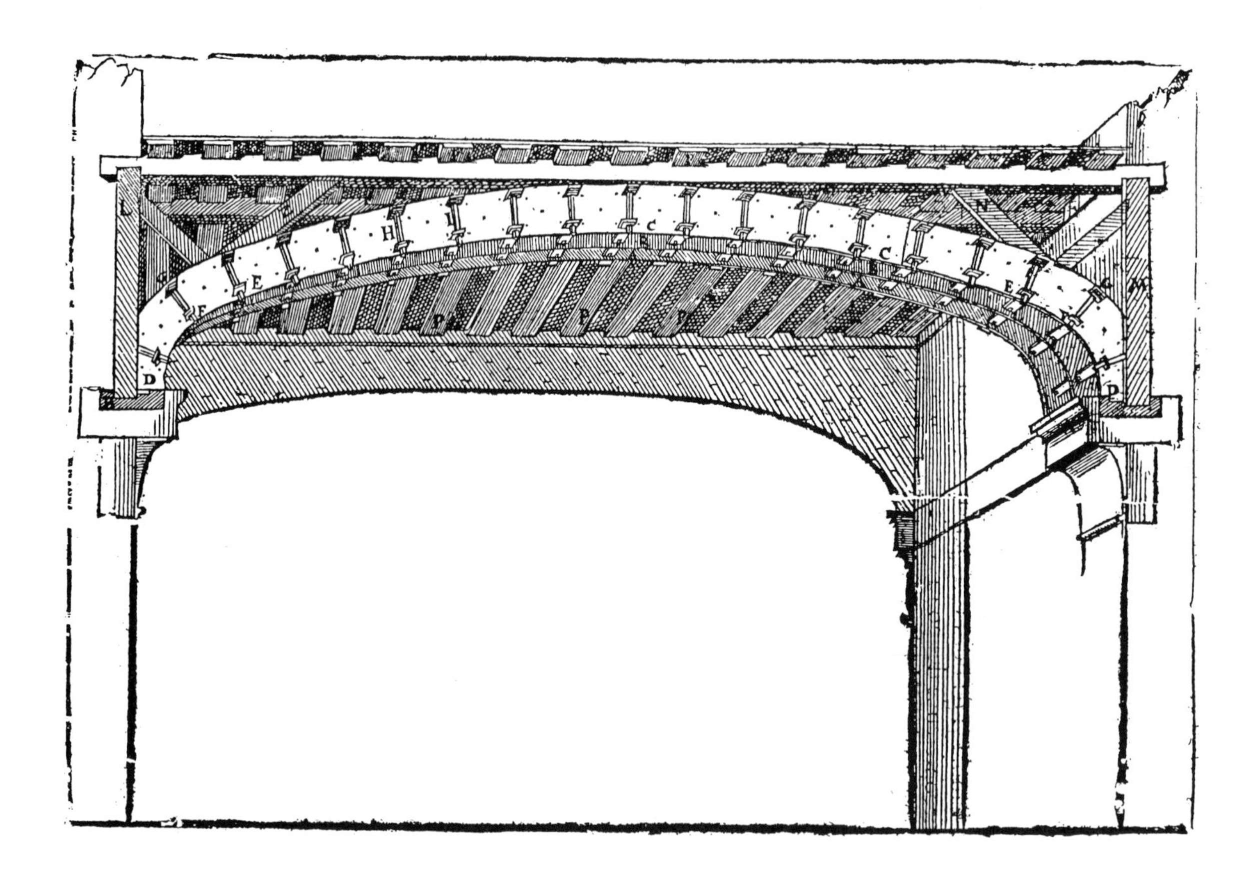

5.36 Philibert de l'Orme: segmented timber arch for floor support, ca. 1570.

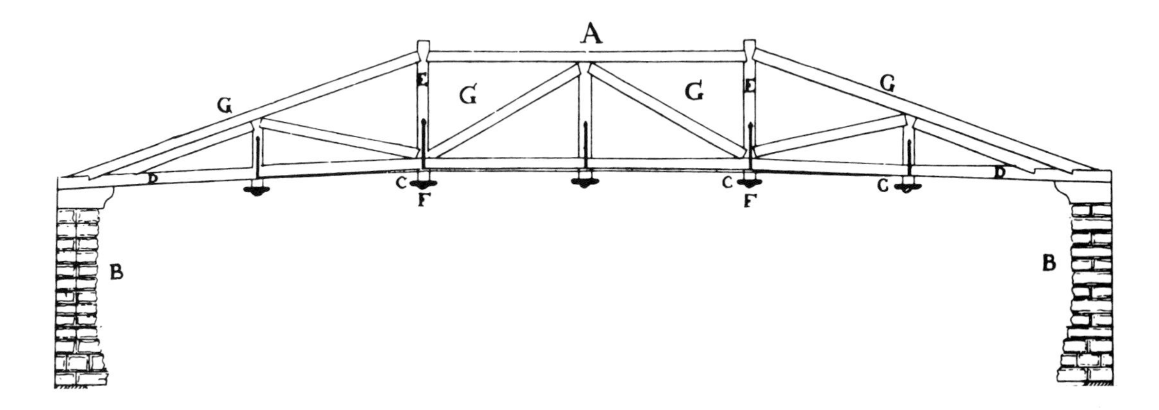

5.37 *Cismone Bridge truss, illustrated by Palladio.*

elements (one example is illustrated in figure 5.36). Although invented mainly for economy, to conserve large timbers and to facilitate the reuse of older construction, or even to utilize the mill ends of timbers of about 25 × 130 cm (Philibert de L'Orme, 29), of nearly equal importance was his desire to obviate the limitation in span dictated by the availability of long timbers. Philibert envisioned arched spans of 60 meters and large domical vaults of ogee profile. As with any arch, however, such construction would necessarily place appreciable lateral thrust on a supporting wall, so the Philibert system was not practical for use in tall buildings.

Modern-looking trusses for long-span application are described by Andrea Palladio in his *I Quattro Libri dell' Architettura*, first published in 1570 (Palladio, III. 7.). By using bolted metal cramps to join the vertical elements to the lower chord, the bridge illustrated in figure 5.37, a 30-meter span over the Cismone River in northeastern Italy, should have performed quite well. With the major loadings coming from the walkway, all of the diagonals and upper chords are in compression, while all the verticals and lower chords are in tension; and these forces are appropriately taken into account in the joining. In the same spirit that led to the lateral, wind bracing of medieval roof trusses, Palladio also recommends the addition of lateral bracing for longer-span bridge trusses.

The only remaining technical question pertains to the makeup of the lower chords of the bridge truss. Palladio's drawing (figure 5.37) seems to indicate that they are formed from continuous, single timbers spanning the entire length of the bridge. But as discussed above in terms of size of available European timber, this is most unlikely. Some type of tension splice, possibly reinforced with iron strapping, must have been used at several locations in the lower chords.

One of the largest-span roofs of this period was designed for the first major building project of Christopher Wren, the Sheldonian Theater at Oxford. Wren based his design (figure 5.38) on the Roman Theater of Marcellus (as depicted in editions he had seen of Serlio and Vitruvius). But while the Roman semicircular arena-theater was covered only by a canvas awning, Wren had to provide permanent closure because of the English climate and also to allow the building to be used by the Oxford Univer-

5.38 *Sheldonian Theatre, Oxford, 1662–1666.*

sity Press, whose books were to be stored in the attic above the auditorium.

Wren's design is notable on two accounts: it was the first building to depart from the Gothic tradition at Oxford, and it incorporated an unusually long-span ceiling over the auditorium without any supporting columns. Instead of using a more efficient triangulated truss for the ceiling, Wren chose a type of *tied arch* where the elements of the upper chord act more or less in pure compression (like the arch illustrated in figure 3.14), while the lower chord acts mainly as a tension tie, relieving the walls of outward thrust. Wren's drawing (figure 5.39) indicates that the lower chord was composed of metal-fastened, scarf-jointed elements. Structural modeling of the Sheldonian roof truss indicated that distress would not have been expected from the level of stresses caused by normal loadings (Dorn and Mark, 163). Perhaps, then, the weight of stored books in the attic was abnormally large and unevenly distributed (recall that an arch performs best under uniform loading), for the original trusses were replaced in 1802.

In this wide-ranging survey, it is clear that roofs represent a major economic undertaking involving many craftsmen—carpenters, smiths, and woodsmen—as well as resources in timber, metals, and terra cotta. The invention of the tied truss was one of the most important developments in the history of architecture. Like foundations, they represent a sizable proportion of the production of large buildings, and clearly, the stylistic and technological virtuosity displayed in their design should no longer be overlooked.

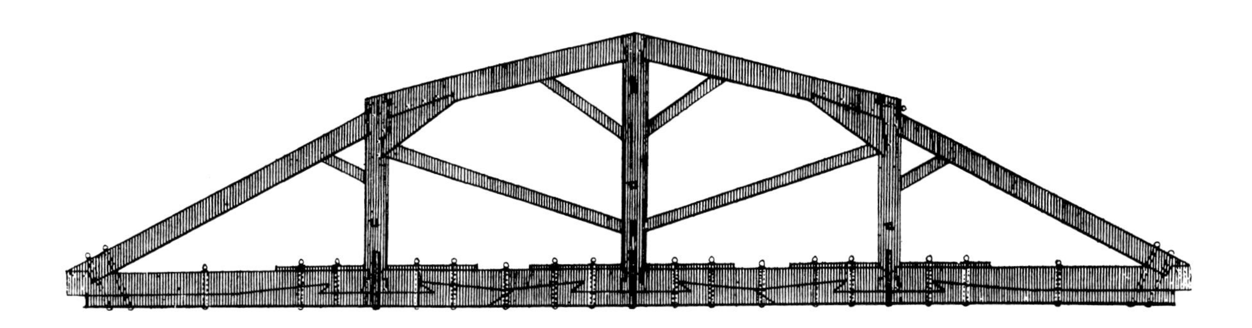

5.39 *Sheldonian Theatre: roof framing for support of the auditorium ceiling (after Wren).*

Bibliography

Adam, Jean-Pierre. *La Construction Romane, Materiaux et Techniques*. Paris, 1984.

Alberti, Leon Battista. *On the Art of Building*, trans. J. Rykwert, N. Leach, and R. Tavenor. Cambridge, MA, 1988.

Alcock, N. W. *Cruck construction: an introduction and catalogue*. CBA Research Reports, 42, 1981.

Baillie, M. G. L., *Tree Ring Dating and Archaeology*. Chicago, 1982.

Baines, Frank, *Report on the Roof Condition of Westminster Hall*, Commissioned Document 7436, House of Commons, 1914.

Bohland, J. "Das konstruktive Gefüge der Holzdecke von St. Michael." *Niedersachsiche Denkmalpflege*, 2 (Hildesheim, 1957), pp. 19–25.

Braunfels, Wolfgang, *Monasteries of Western Europe*, trans. A. Laing. London, 1972.

Brunskill, R. W. *Timber Building in Britain*. London, 1985.

Charles, W. F. B. *Medieval Cruck-Building and its Derivatives*. London, 1967.

Choisy, A. *L'Art de bâtir chez les Romains*. Paris, 1873.

Cichorius, Conrad. *Die reliefs der Traianssäule*, 2 vols., 2 atlases. Berlin, 1896–1900.

Cordingly, R. A. "British Historical Roof-Types and their Members." *Transactions of the Ancient Monuments Society*, New Ser. IX (1961), pp. 73–118.

Coulton, J. J. *Greek Architects at Work: Problems of Structure and Design*. London, 1977.

Coulton, J. J. *The Architectural Development of the Greek Stoa*. Oxford, 1976.

Courtenay, L. T. "Westminster Hall and Its Fourteenth-Century Sources." *Journal of the Society of Architectural Historians*, 43 (December 1984), pp. 295–309.

Courtenay, L. T. "Viollet-le-Duc et la flèche de Notre-Dame de Paris: la charpente gothique au XIII et XIXe siecles." *Journal d'histoire de l'architecture*, 2 (1989), pp. 53–68.

Courtenay, L. T. "The Westminster Hall Roof: A New Archaeological Source." *British Archaeological Association Journal*, 143 (1990), pp. 95–111.

Courtenay, L. T. and R. Mark. "The Westminster Hall Roof: An Historiographic and Structural Study." *JSAH*, 46, (December 1987), pp. 374–393.

Crook, John. "The Pilgrim's Hall, Winchester: Hammerbeams, Base Crucks and Aisle-Derivative Roof Structures." *Archaeologia*, 109 (1991), pp. 129–159.

Currie, C. R. J. "A Romanesque Roof at Odda's Chapel, Deerhurst, Gloucestershire." *The Antiquaries Journal* (1989), pp. 58–66.

de Fine Licht, K. *The Rotunda in Rome: A Study of Hadrian's Pantheon.* Copenhagen, 1968.

Deneux, Henri. *Charpentes,* Centre de Recherches sur les Monuments Historiques, vols. I and II. Paris, 1959–1960.

Deneux, Henri. "Evolution des Charpentes du xie au xviiie siècle." *L'Architect,* 4 (Paris, 1927), pp. 49–68.

Dio, Cassius Cocceianus. Loeb Edition, Cambridge, 1914–1927.

Dorn, Harold, and R. Mark. "The Architecture of Christopher Wren." *Scientific American,* 245 (July 1981), pp. 160–173.

Durand, Georges. *Monographie de l'église cathédrale Notre-Dame d'Amiens,* 2 vols. Amiens, 1901–1903.

Fletcher, J. M., and P. S. Spokes. "Origin and Development of Crown-post Roofs." *Medieval Archaeology,* VIII (1964), pp. 152–183.

Forbes, Robert J. *Studies in Ancient Technology.* Leiden, 1964.

Forsyth, George H. "The Monastery of St. Catherine at Mount Sinai: The Church and Fortress of Justinian." *Dumbarton Oaks Papers,* 22 (1968), pp. 3–19.

Forsyth, George H. *Sinai and the Monastery of St. Catherine.* New York, 1980.

Genicot, L. "Charpentes du xie au xixe siècle en Wallonie." *Bulletin: Commission royale des Monuments et des Sites,* IV (1974), pp. 29–51.

Ginouvès, René, and R. Martin. *Dictionnaire methodique de l'architecture greque et romaine,* I, *Matériaux Techniques de Construction.* Rome, 1985.

Goodburn, D. M. "The Earliest Timber Framed Building in Britain?" *Rescue Archaeology,* 5 (1990), 8.

Hauglid, Roar. "The Trussed-Rafter Construction of The Stave Churches in Norway." *Acta Archaeologica,* 43 (1972), pp. 19–55.

Hewett, Cecil. *English Cathedral Carpentry.* London, 1974.

Hollstein, E. *Mitteleuropaische Eichenchronologie.* Mainz am Rhein, 1979.

Horn, Walter. "On the Origins of the Medieval Bay System." *JSAH,* 17, (January 1958), pp. 2–23.

Janse, Hans. *Houten kappoen in Nederland 1000–1940,* Delft, 1989.

Kuniholm, P. I., and C. L. Striker. "Dendrochronological Investigations in the Aegean and Neighboring Regions 1983–1986." *Journal of Field Archaeology,* 14 (1987), pp. 385–398.

Kraft, Jean Charles. *Traité sur l'art de la charpente theorique et pratique,* 5 vols. Paris, 1819.

Krautheimer, R., et al. *Corpus Basilicarum Christianarum Romae,* I-IV. Vatican City and New York, 1937–1977.

Lange, Konrad von. *Basilica at Pompei.* Leipzig, 1885.

Lemaire, Raymond. *Les Origines Du Style Gothique en Brabant.* Brussels, 1906.

MacDonald, William L. *The Pantheon: Design, Meaning and Progeny.* Cambridge, MA, 1976.

Martin, Rowland. *Manuel D'Architecture Grècque:* I, *Matériaux et Techniques.* Paris, 1965.

McClendon, C. B. *The Imperial Abbey of Farfa.* New Haven, 1987.

Meiggs, R. *Trees and Timber in the Ancient Mediterranean World.* Oxford, 1982.

Meirion-Jones, G. I. "The Vernacular Architecture of France: An Assessment." *Vernacular Architecture,* 16 (1985), pp. 1–17.

Moles, Antoine. *Histoire des Charpentes.* Paris, 1949.

Ostendorf, F. *Geschichte des Dachwerks.* Berlin and Leipzig, 1908.

Oswald, F. "Römische Basilika und ottonische Kirche, St.Peter auf der Zitadelle in Metz." *Fruhmittelalterliche Studien,* I, Trier (1967), pp. 156–169.

Panofsky, Erwin. *Abbot Suger: on the Abbey Church of St. Denis and its Art Treasures,* 2d ed. Princeton, 1979.

Palladio, Andrea. *The Four Books of Architecture,* trans. Isaac Ware, London, 1738; reprint, New York, 1965.

Philibert de L'Orme. *Nouvelles inventions pour bien bastir et a petits fraiz* [1567]. Pérouse de Montclos, Jean-Marie, ed. Paris, 1988.

Rackham, Oliver. *Trees and Woodland of the British Landscape.* London, 1976.

Rondolet, J. *Traite theorique et pratique de l'art de batir,* 6 vols. Paris, 1802–1817.

Saalman, Howard. *Filippo Brunelleschi: The Cupola of Santa Maria del Fiore.* London, 1980.

Sackur, Walter. *Vitruv, Technik und Literature.* Berlin, 1925.

Salzman, L. F. *Building in England.* Oxford, 1952.

Simpson, Gavin W. Field report, in press.

Smith, John T. "The Early Development of Timber Buildings: The Passing Brace and Reversed Assembly." *Architecture Journal,* 131 (1974), pp. 238–263.

Smith, John T. "Norwegian Stave-church Roofs from an English Standpoint." *Universitetets Oldsaksamling Arbok 1975/1976.* Oslo, 1977, pp. 123–136.

Smith, John T. "Medieval Roofs: a classification," *Archaeological Journal*, 115 (1958), pp. 111–149.

Stevens, G. P. "Remarks on the Chryselephantine Statue of Athena." *Hesperia*, 24 (1955), 20–27.

Theophrastus. *Enquiry Into Plants*, trans. A. Hort, 2 vols. Cambridge and London, 1968.

Trevor-Hodge, A. *The Woodwork of Greek Roofs*. Cambridge, 1960.

Viollet-le-Duc, E. E. *Dictionnaire raisonné de l'architecture française du XIe au XVIe siècle*, 10 vols. Paris, 1854–1868: see especially "Charpente," vol. 3.

Vitruvius, *De Architectura*, ed. Cesare Cesarino, Como, 1521; *The Ten Books of Architecture*, trans. H. Morgan, Cambridge, 1914.

Wren, Christopher. *Parentalia*. London, 1965; reprint, 1750.

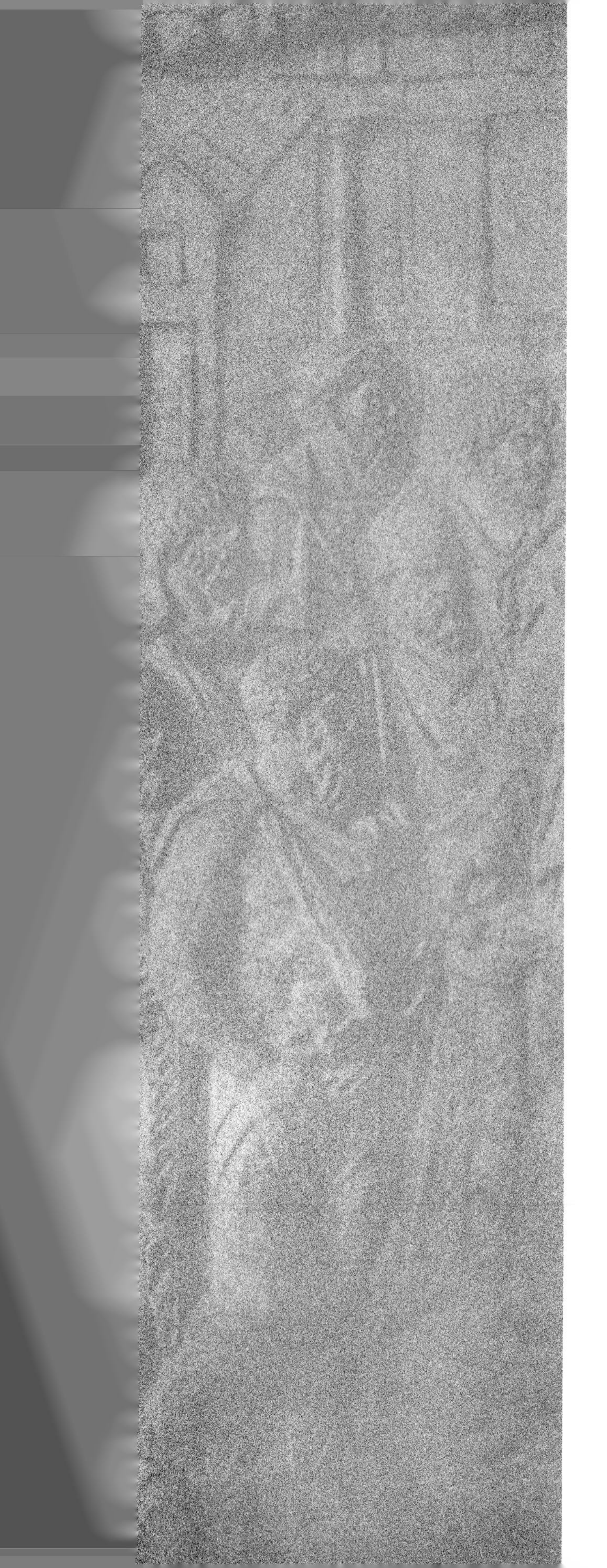

By focusing on a wide range of historic buildings, from their below-ground-level foundations to the peak of their timber roofs, the typical developmental pattern of prescientific technology becomes evident. Much of what was then learned through experience and observation is available to today's designers from scientific analysis. Earlier buildings effectively served as "models" for new design. This process is manifest even in "breakthrough" building projects such as Hadrian's Pantheon, Justinian's Hagia Sophia, Suger's St. Denis, Herland's roof for Westminster Hall, and Wren's dome for St. Paul's. Even though the Pantheon's span is almost twice that of known earlier domed buildings, all the necessary constructional techniques for this form of structure were well in hand before its erection was undertaken. Similarly, analysis of the structure of Hagia Sophia shows that its designers possessed no greater understanding of structures than might have come from building SS. Sergius and Bacchus in Constantinople or the Pantheon and the Minerva Medica in Rome. The choir of St. Denis was undoubtedly an outgrowth of experimental church building prevalent in the Île-de-France in the first half of the twelfth century, just as the hammer-beam roof of Westminster Hall was a product of the English Late Gothic tradition of monumental carpentry. And Christopher Wren's design for the dome of St. Paul's owed at least as much to earlier dome projects, going back as far as the thirteenth-century additions to St. Mark's, as it did to seventeenth-century science.

The elegance of these structural solutions points to other factors contributing to the successes of the master builders. Steps taken to eliminate undesirable behavior of the building fabric observed during the process of construction—particularly cracking in the fabric, but sometimes excessive deformation—led to refinements in design. In a sense, this design strategy was a precursor to today's electronic instrumentation used to investigate building behavior, on the basis of which necessary steps are taken to modify inadequate structure.[1] At the same time, it has to be acknowledged that some further help is offered by the forgiving nature of typical masonry construction. Small deviations in geometry, caused by errors in building or resulting from large structural deformations, have much less effect on the stability of masonry than do deviations of similar scale in modern construction incorporating far more slender elements of steel or reinforced concrete. Even though masonry is extremely weak in tension, the presence of highly localized regions containing tension rarely causes the kind of catastrophic collapse experienced at Beauvais (chapter 3).

All of these factors leading to design success, except for the last (related to material behavior alone), are predicated on an intimate relationship between the process of design and of building. We know, for example, that the great Gothic cathedrals were constructed by teams of skilled masons and carpenters and—most important for maintaining the continuity of experience to new building—apprentices who rose through the ranks to become master builders. But this process came to a fitful end across most of Europe during the Renaissance of the fifteenth and sixteenth centuries. Master builders were displaced as principal designers of monumental buildings by artist-architects typified by the goldsmith Brunelleschi, the writer Leon Battista Alberti (1404–1472), and the sculptor Michelangelo. As Catherine Wilkinson notes, "this tradition was founded [in central Italy] on the belief (well rooted by the fifteenth century) that any artist could design a building since it was the conception of the work that mattered rather than the construction."[2]

An indication of this change is the bitter dispute that erupted between Gian Lorenzo Bernini (1598–1680) and Francesco Borromini (1599–1667) over the structural problems arising from the two *campanili,* or bell towers, Bernini had designed for the facade of St. Peter's in Rome. Bernini, trained as a sculptor, with little practical experience in building, was also a talented poet and scholar, which facilitated his entry as a favored member into the humanist circle in the Vatican led by Cardinal Scipione Borghese (later Pope Urban VIII). His many commissions as both sculptor and architect consequently owed a great deal to this association. Borromini, on the other hand, came out of the older artisan tradition and had worked his way up from stonecutter to draughtsman in the workshop of his mentor Carlo Maderno, at that time the architect in charge of the work at St. Peter's.

It was Bernini and not Borromini, however, who succeeded Maderno as architect of St. Peter's. Failing to appreciate the soil conditions at the site, Bernini encountered severe problems with the bell towers framing the ends of Maderno's facade that he began in 1638. Excessive settlement of the south tower soon threatened the stability of the entire facade, and Borromini launched a bitter attack on his rival, accusing him of a dangerous lack of practical building experience. Although his criticism interrupted Bernini's architectural career for almost a decade, Borromini was not chosen to replace Bernini at St. Peter's. And despite the necessary demolition of both towers in 1647, it became clear that Bernini's intellectual approach to architectural design was to be the wave of the future. Borromini's practical knowledge of construction, developed by years of experience, afforded him no advantage over his humanistically educated rival.

Facilitated by the development of printing, translations of the major Italian architectural treatises of the Renaissance became accessible throughout Europe by the seventeenth century. These also helped to set the builder apart from the architect, who came to have a cultural/social background similar to that of his patrons. Thus the gap between designer and builder widened, and in addition, the publication of drawings of existing buildings, accompanied by measurements and in some cases empirical rules for structure, tended to codify design.[3] Once certain building types were established and canons accepted for construction, there was little incentive for modification until the pressing needs of nineteenth-century industrial development and the introduction of new construction materials brought the more common use of scientific methods into building design. Indeed, there is consensus among historians that until the nineteenth century, the level of structural experimentation in European architecture never again approached its Gothic zenith.

On the other hand, the new professionalism of architects hastened another exceedingly important development. Exceptional practitioners such as the Sienese Francesco di Giorgio (1439–1501), the Parisian Philibert de l'Orme, and the Vicentine Andrea Palladio, though beginning their careers as masons or stonecutters, readily adapted to the new social status of architects and authors. They and other architects produced elaborately illustrated books, often with dimensioned drawings, aimed at patrons and general consumers of humanist culture. It was these books, according to Pamela Long, that helped initiate European belief in the value of combining theory and practice (or experiment) and resulted in the modern scientific method. "The significance of architectural writings is not that they were the first result of the interaction of humanist and artisan culture [note Galileo's remarks about the Venetian artisans, quoted in chapter 1], but they constituted the first textual tradition which, as a result of that interaction, disseminated in writing the ideal of the unity of theory and practice. The explicit and repeated articulation of that ideal in widely disseminated treatises and commentaries had important consequences for the intellectual history of the period. . . . Francis Bacon [1561–1626] believed that he had 'established forever a true and lawful marriage between the empirical and the rational faculty.' Actually, the marriage of theory and practice had occurred in the architectural tradition of the two preceding centuries."[4] The genesis of that tradition is to be found in the two-millennia span of architectural technology described in this book.

Notes

1. Robert Mark, *Light, Wind, and Structure* (Cambridge/New York, 1990), p. 170.

2. Catherine Wilkinson, "The New Professionalism in the Renaissance," S. Kostof, ed., *The Architect: Chapters in the History of the Profession* (New York), 1977, p. 134.

3. L. Shelby, and R. Mark, "Late Gothic Structural Design in the 'Instructions' of Lorenz Lechler," *Architectura,* 9 (1979), 113–131.

4. Pamela O. Long, "The contribution of architectural writers to a 'scientific' outlook in the fifteenth and sixteenth centuries," *Journal of Medieval and Renaissance Studies,* 15:2 (Fall 1985), p. 298.

AEDICULA
A canopied niche for a statue or an image.

APSE
Semicircular or polygonal termination of the eastern end of a church.

ARCADE
A series of arches on columns or piers; the arcaded lower story of a church.

ASHLAR
Square-hewn dressed stone; also, masonry in which all stones are squared, giving a uniform pattern of vertical and horizontal joints.

ARCH
A curved structural member, generally carrying a distributed load transverse to a line drawn between its ends, or *springing*, producing principally internal compression. The *keystone* is the uppermost voussoir of a masonry arch. A *diaphragm arch* is a transverse, thin wall supported by an arch, usually high over a nave. A *corbel arch* is not a true arch, but rather composed of masonry elements projecting slightly past the elements beneath into an opening. A *relieving arch* is an arch within a wall designed to relieve the wall below from the weight of the wall above. A *quadrant arch* is an internal buttress similar both in form and function to an external flying buttress.

ARCUATED
Construction based on the arch.

AULA
In ancient Greek architecture, a hall or court.

BASILICA
Of Roman origin, a large rectangular hall with a high central space, flanked by lower side aisles. A *basilican plan,* therefore, usually indicates three parallel rectangular aisles.

BATTEN
A light strip of wood used over a seam between boards.

BAY
Compartments into which a building is divided, normally marked by vertical supports.

BEAM
Generally, a slender structural member carrying loadings transverse to its longitudinal axis, producing internal shear and bending. A *cantilever beam* is supported at only one end. A *hammer beam* is a short horizontal timber member projecting inward from a wall and usually braced from below.

BLADES
See cruck.

BUTTRESS
A massive upright structure, usually of masonry, that resists lateral, overturning forces (such as those transmitted by flying buttresses). An *intermediate buttress* is the slender upright support to the center of the flying buttress in a five-aisled Gothic church; a *wall buttress* is a projection attached to a wall to enhance stability. A *flying buttress* is an arch-like external structure that normally acts as a compression brace against lateral forces; and a *spur* is a lateral wall under a side-aisle roof that acts structurally in the same manner as a flying buttress.

CAPITAL
The transitional block above a column or pier.

CATENARY
The form assumed by a hanging chain under the action of gravity, normally very close to parabolic.

CELLA
The inner, enclosed room of an ancient temple.

CENTER OF GRAVITY
A point, normally within a volume, about which the volume is balanced; hence the weight of the volume can be taken as acting at the center of gravity.

CENTERING
Temporary shoring, usually constructed of timber, to support arches or vaulting during construction. Rather than being built up from ground level, *flying centering* is usually supported by corbels near the springing of an arch or vault.

CHOIR
The portion of a church where services are sung, generally at the eastern end.

CLERESTORY
The wall, or story, that rises above the aisle roof, pierced by windows.

COFFER
A recessed panel in a ceiling or on the underside of a dome or vault.

COMPRESSION
An axial pushing force or stress tending to shorten a structural member.

CONCRETE

An artificial stone composed of cement, water, and *aggregates* such as crushed stone and sand. Concretes made with *hydraulic cement,* a cement that combines chemically with water, can be used for underwater construction.

COLLAR

A horizontal timber connecting pairs of rafters at an elevation between their feet and their apex.

COLONNETTE

A light column, not usually an element of a building's primary structure.

COURSE

A line of stone blocks. *Coursed* masonry is composed of lines of blocks, but it may contain *broken courses* when the joints are not all in the same plane.

CROSSING

The space formed by the central piers at the intersection of the longitudinal and lateral spaces in a cruciform-plan church.

CRUCK

A pair of straight or curved timbers (*blades*) that serve both as wall posts and rafters.

DEAD WEIGHT

The self-weight of a structure.

DENDROCHRONOLOGY

Dating of timber by comparing specimen tree-ring patterns with an established long-time compilation of the region known as a *master chronology.*

DOME

Usually an axisymmetric structure of arched section forming a ceiling or roof. A *semidome* is half a *hemispherical dome* or a dome having a semicircular cross section. *Step rings* are the concentric, projecting rings often placed around the base of classical, or classically derived, domes.

DRUM

A cylindrical or polygonal wall supporting a dome or a lantern.

ENTABLATURE

In classical architecture, the horizontal group of members carried over the exterior columns, comprising (from bottom to top) the: *architrave, frieze,* and *cornice.* The *triglyphs* and *metopes* are alternating decorated panels on the surface of the Doric frieze.

EXEDRA

A large, semicircular extension of a building.

FLÈCHE

A slender wooden spire rising from a roof, often covered in lead.

FOOTING

The projecting base of a pier or wall that distributes loadings to the subsoil. *Continuous wall footings* are a continuous projection below a wall, and *single column footings* are isolated structures below a pier.

FRAME
A skeletal load-bearing structure. A *rigid frame* is a planar assembly of beam-like elements rigidly connected to each other, generally formed of timber in early construction, and of steel or reinforced concrete in large-scale modern construction.

GALLERY
In a church, the roofed or vaulted space over the side aisle forming a second story above the arcade story.

HEMICYCLE
The semicircular structure of the rounded termination of a church.

LAGGING
Wood planks used for the temporary support of the masonry of an arch or vault in construction.

LANTERN
An open-walled tower on the top of a dome or roof to admit light to the space below.

LIGHT INTENSITY
The measure of light emission from a natural or artificial source. *Illuminance* is the measure of surface brightness. *Light-path length* is the distance from a source to an illuminated surface.

MERIDIAN
A radial generating line of a dome surface.

MODELING
Modeling techniques are used in engineering to find deformations and force and stress distributions in complex structures. *Physical modeling* involves taking measurements, usually from a small-scale model, which are *scaled* to give the response of a *prototype*, full-scale structure. With *photoelastic modeling*, these measurements are derived from optical interference patterns using polarized light. With *numerical modeling*, the abstracted material properties, geometry and loading of the prototype are programmed into an electronic computer; in *finite-element* (numerical) modeling, the prototype form is represented by a series of coordinates taken at (finite) intervals on its surface.

MULLION
A light, upright division member within windows.

NARTHEX
The vestibule of a church.

NAVE
The main portion or central aisle of a church or basilica.

NOGGING
A wall filling between wood studs, usually of brick or clay.

ORTHOGONAL PROJECTIONS
The (horizontal) plan, (vertical) elevation, and section drawings taken through a building.

ORTHOSTATS
Long blocks of stone, laid horizontally, upon which the column rows of ancient temples usually rest.

PEDIMENT
The triangular face of a roof gable.

PENDENTIVE
The triangular segment of vaulting at the base of a dome used for transition from a round- to a square-planned space.

PIER
An upright structure of masonry acting mainly to support vertically acting loads.

PILE
A heavy timber driven into soil, sometimes under water, either to translate building loads through weak soils to stronger soils beneath, or through skin friction to the surrounding soil.

PINNACLE
A small spire, usually surmounting a buttress or roof.

PLATE
A longitudinal timber set on top of a foundation or wall. An *arcade plate* is a longitudinal timber placed on top of wall posts in aisled construction. A *flying plate* is a longitudinal timber set in from the plane of the wall and carried by extended sole pieces (horizontal timbers set above and perpendicular to the wall plates) and/or brackets.

PURLIN
A longitudinal timber set in the plane of the roof to support common rafters.

PUTLOG HOLE
A blind opening in a wall, usually square, used for the support of scaffolding.

RADIOCARBON DATING
An experimental dating technique based on measuring the decay of the atmospheric radiocarbon assimilated by living plants.

REACTIONS
The internal forces acting at the supports of a loaded structure.

RAFTER
The sloping roof beam, from the ridge of the roof to the wall. A *principal rafter* provides primary support to longitudinal purlins that in turn support the lighter *common rafters* of the roof.

ROTUNDA
A circular hall usually topped by a dome.

SCANTLING
The dimensions of a timber, especially of its cross-section.

SHEAR
A force or stress acting transverse to the axis of a structural member tending to cause sliding between its constituent elements.

SOILS
Predominant soils include sands, silts, and clays, identified according to descending order of grain size. *Soil bearing capacity* is the maximum pressure that can be developed under a footing without leading to damaging, large movement. *Soil permeability* is the measure of the rate at which water flows through soils. *Soil shear strength* is a measure of resistance to sliding between soil grains.

SPANDREL
The triangular wall areas over the haunches of an arch or, in multistory buildings, the area of wall between the vertical supports and upper and lower windows.

SPUR WALL
The transverse, triangular wall above the side aisle of a church that provides lateral support to the piers as well as to the side-aisle roofs at each bay.

SQUINCH
An arch, or corbeling, at the base of a dome or cloister vault used for transition to a square-planned space below.

STABILITY
The capacity of a structure to remain in stable equi librium under the action of applied forces.

STOA
A long, sheltered, colonnaded porch.

STIFFNESS
A measure of a structure's resistance to deformation.

STRAIN
A measure of local deformation within a structure.

STRENGTH
The level of stress that causes a material to fail.

STRESS
A measure of the local intensity of internal force within a structure.

STRUT
A slender structural member that resists compressive force along its axis. A *raking strut* is an inclined compression member (often used in pairs) set at an angle to the horizontal tie beam.

STYLOBATE
The base of a classical temple providing support for the columns.

SURCHARGE
The mass placed over the haunches of an arch or vault, or a pendentive, to enhance structural stability.

TALUS
The enlarged, sloped base of a fortified wall.

TENSION
An axial stretching force or stress tending to elongate a structural member.

TEMPLATE
A pattern used to establish the profile of cut stone—used also to profile photoelastic models.

THERMOLUMINESCENCE DATING
An experimental dating technique based on measurements of natural radioisotopes in fired clay.

TIE BEAM
A main horizontal member of a timber roof truss that acts in tension to prevent roof rafters from spreading.

TIMBER
The constructional wood from a tree generally of at least 60 cm in girth, which yields a heartwood at least 15 cm square.

TRABEATED
Post and lintel construction.

TRACERY
Ornamental mullions of stone set into medieval windows.

TRACING FLOOR
The flat surfaces used for incising the form of full-scale building elements.

TRANSEPT
The transverse arms of a cruciform-plan church, usually forming a separation between the choir and the nave.

TRANSTRA
In Roman construction, horizontal roof timbers.

TRIFORIUM
The wall passage above the arcade story and below the clerestory in mature Gothic churches.

TRUSS
An assembly of struts and ties forming triangles and carrying loadings applied transverse to its horizontal axis. The primary horizontal members of a truss are *chords*.

TYMPANUM
The recessed surface below a pediment or arch.

VAULT
A ceiling or roof structure of arched section. A *barrel vault* is a continuous vault of semicircular section. *Groined vaults* are formed by intersecting two orthogonal barrel vaults, the *groins* being the lines of intersection. A *cloister vault* is formed by intersecting singularly curved surfaces ("sails") above a polygonal base; it is similar to a dome in form and structure. *Quadripartite,* ribbed groin vaulting divides each bay into four compartments; *sexpartite* vaulting divides each pair of bays into six compartments. *Tierceron vaults* exhibit extra ribs, making the plan of the vault appear star-shaped; *lierne vaults* deploy shorter ribs linking diagonal and tierceron ribs. *Fan vaulting* forms conoid surfaces of rotation about the axes of supporting piers from an array of uniform ribs. *Ribs* are the arches that project below the vault surface. The *webbing* is the stone surface of a vault seen as infilling between the ribs.

VOUSSOIR
Wedge-shaped stone used as a building block of a masonry arch or vault.

Index

Photo Credits

Alinari/Art Resource, N.Y., 3.51; J. Austin, 3.67; C. Cichorius, 5.3, 5.4, 5.13, 5.14; L. T. Courtenay, 3.3, 3.21, 3.41, 3.53, 5.1, 5.10, 5.21, 5.26, 5.27, 5.29, 5.33; S. Ćurčić, 2.9, 3.24, 3.45, 4.16; M. Davis, 1.7; W. H. Fox-Talbot, Science Museum London, 3.34; J. Herschman, 3.56, 3.59, 3.65, 3.68; F. R. Horlbeck, 4.20, 4.27; V. Jansen, 3.70; Le Secq, 3.85; C. Maines, 2.25, 3.52; Foto Marburg/Art Resource, N.Y., 3.54, 4.15; Foto Marburg/W. Horn, 5.8; R. Mark, 2.1, 3.2, 3.7, 3.15, 3.16, 3.20, 3.30, 3.35, 3.38, 3.40, 3.46, 3.47, 3.57, 3.71, 3.72, 3.77, 3.78, 3.80, 3.87, 3.88, 4.7, 4.8, 4.19, 4.21, 4.30, 4.37, 5.38; Medieval Academy, 3.60; Monuments Historiques, Paris, 2.27; H. H. Richardson collection, GSDL, Loeb Library, Harvard University, 3.18, 4.31; E. C. Robison, 2.3, 2.10, 4.22, 4.29; R.C.H.M., 5.30; SPADEM, 1.5, 4.18; SPADEM/C. Méstral, 3.82; J. W. Williams, Jr., 3.1, 3.49, 4.11.